Biomass and Carbon Fuels in Metallurgy

Assoc. Prof. Ing. Jaroslav Legemza, PhD

Prof. Ing. Mária Fröhlichová, PhD

Assoc. Prof. Ing. Róbert Findorák, PhD

CISP

CRC Press
Taylor & Francis Group
Boca Raton London New York

CRC Press is an imprint of the
Taylor & Francis Group, an **informa** business

Some parts of this monograph were written with the support of the Slovak Research and Development Agency, Ministry of Education, Science, Research and Sport of the SR.

Translation:
Assoc. Prof. RNDr. Pavol PALFY, PhD

Language correction:
Jozef Veselský, translator/interpreter, SAPT member

Reviewers:
Prof. Ing. Ivan Imriš, Dr
Prof. Ing. Ľubomír Mihok, PhD
Assoc. Prof. Ing. Ján Kret, PhD

CRC Press
Taylor & Francis Group
6000 Broken Sound Parkway NW, Suite 300
Boca Raton, FL 33487-2742

First issued in paperback 2021

© 2020 by CISP
CRC Press is an imprint of Taylor & Francis Group, an Informa business

No claim to original U.S. Government works

ISBN 13: 978-0-367-22242-0 (hbk)
ISBN 13: 978-1-03-223830-2 (pbk)

Visit the Taylor & Francis Web site at
http://www.taylorandfrancis.com

and the CRC Press Web site at
http://www.crcpress.com

Foreword (preface)

The publication "Biomass and Carbon Fuels in Metallurgy" was produced at the Department of Ferrous Metallurgy and Foundry, Institute of Metallurgy, Faculty of Materials, Metallurgy and Recycling, Technical University of Košice, during the years 2012–2017. The knowledge and information contained herein originate from various research projects implemented by authors in the course of many years, as well as from professional and popularisation resources published in Slovakia and the world.

The idea of the publication "Biomass and Carbon Fuels in Metallurgy" arose due to the current need of the professional and lay public for information relating to the use of carbonaceous (traditional and alternative) fuels in the metallurgical sector. The authors have attempted to provide both the latest information and experience in the presented field (What is new?) and also to predict the prospects of the area in the future (What will be needed? What has a future?).

This publication would hardly be completed without the support of research, teaching and operational activities of many researchers, teachers and representatives of the industrial sector, who have been collaborating with the authors of the publication for a long time. It is almost impossible to list everyone here (some already deceased), but the authors wish to thank at least these colleagues: Š. Majerčák, A. Majerčáková, Ľ. Mihok, A. Kucková, D. Baricová, A. Markotic, J. Kret, D. Jakubéczyová, A. Mašlejová, P. Ballok, A. Trišč, M. Olejár, P. Nemčovský, J. Bernath, J. Čurilla, Kočuta M., J. Kendera, R. Maliňák, G. Lešinský, F. Bakaj, M. Džupková, M. Čižmárová, Z. Semanová, J. Leško, J. Hudák, R. Mežibrický.

The authors thank the representatives of metallurgical enterprises in the Slovak Republic as well as colleagues from the Faculty of Materials, Metallurgy and Recycling, especially the dear colleagues from the Department of Metallurgy and Foundry, Technical University of Košice.

Content

Used abbreviations and important explanations

Biomass – a biogenic organic matter of plants or animals (usually including its secretions/excretions, waste from it, etc.) used as an energy source.

Cellulose $(C_6H_{10}O_5)n$ – polysaccharide consisting of beta-glucose, which is the main building material of plant cell walls.

CO_2 – carbon dioxide is atmospheric gas consisting of two oxygen atoms and one carbon atom. It has low reactivity and is colourless, non-flammable, heavier than air, produced by biological processes, such as respiration and fermentation, and as a combustion product of carbon compounds in the air, released into the atmosphere.

COREX, HISMELT – industrially used processes for pig iron production by so-called smelting reduction by solid reducing agent – an alternative to the blast furnace process.

DRI (direct reduced iron) – direct production of iron from ore.

Energy (W) – in physics, it is the ability of a physical system to perform work; the unit is Joule (J).

EAF – electric arc furnace. It is the essential metallurgical equipment for steelmaking.

Glucose $(C_6H_{12}O_6)$ – D-glucose, which is found in plants as a product of photosynthesis and serves as a reservoir of energy.

HBI (hot briquetted iron) – briquetted iron produced directly from ore.

IGCC (Integrated Gasification Combined Cycle) – coal is not combusted directly but reacts with oxygen and steam producing synthetic gas – "syngas".

BOF – oxygen converter. It is the essential metallurgical equipment for steelmaking.

Coke – fossil fuel. It is actually refined fuel that is industrially produced by the carbonisation of coal without air, in so-called coke oven batteries.

Maceral – microscopically distinguishable component of coal. According to genetic, physical and chemical properties, macerals in coal are classified into three basic groups – vitrinite, liptinite and inertinite.

Methane (CH_4) – the basic hydrocarbon in natural gas.

MIDREX, FASTMET, FIOR, ITmk3 – industrially used processes for the production of directly reduced iron in solid and liquid form by gaseous or solid reducing agent – an alternative to the blast furnace process.

M_{40} and M_{10} – coke strength M40 and coke abrasion M10 – express the ratio of coke with the grain size above 40 mm or below 10 mm after a test in a drum to coarse coke weighed into the drum before the test, given as a percentage.

PAHs – polycyclic aromatic hydrocarbons

Fuel – natural or synthetically produced substance (a combustible with other possible substances) from which heat (light) can be obtained by burning under economically and ecologically (or economically and hygienically) acceptable conditions.

PCBs – polychlorinated biphenyls

PCDDs – polychlorinated dibenzo-p-dioxins

PCDFs – polychlorinated dibenzofurans

PCI – (Pulverised Coal Injection) – coal in the form of fine powder is injected through tuyeres into a blast furnace.

Ash – a solid residue formed by perfect or imperfect combustion of fuel.

Crude ash – inherent and admixed minerals in carbonaceous fuel.

POPs (persistent organic pollutants) – substances remaining in an environment for a long time with a negative impact on animals, including humans.

Volatile combustible matter (VM) – a mixture of high molecular weight organic compounds of C, H, O, N and S.

Reactivity – the ability of carbonaceous fuel to react with carbon dioxide into carbon monoxide. It depends on the texture arrangement, porosity, ash content and degree of coalification.

Higher calorific value (HCV) – the amount of heat released by the complete combustion of 1 kg of fuel into CO_2, SO_2 and liquid water (H_2O); the unit is $J.kg^{-1}$.

Lower calorific value (LCV, CV), – the amount of heat released under the same conditions, with the difference that water vapour ($H_2O_{(g)}$) is released instead of liquid water; the unit is $J.kg^{-1}$.

TOC (Total Organic Carbon) – represents the total amount of carbon bound in organic compounds, water and gases.

SPM (solid particulate matter) – solid contaminants (in air)

Coal – a fossil fuel. It is a black or brownish-black rock of flammable caustobiolites group, formed in coal basins by processes of coalification from dead plant bodies by the effects of pressure and temperature.

VOC (Volatile Organic Compounds) – non-methane volatile organic compounds

BF – blast furnace is the essential metallurgical equipment for the production of pig iron.

p.i. – pig iron

1. Introduction to the field

The publication "Biomass and Carbon Fuels in Metallurgy" brings contemporary but also new insights into the use of carbonaceous fuels in the metallurgical sector. These fuels are deficient raw materials nowadays. Therefore, it is necessary to find their optimal use. Dwindling traditional fossil fuels and the dynamic growth of the world population are critical factors in our society concerning maintaining the development of industrial production and ensuring energy security. Increasing demand of countries for sufficient energy supply and associated threat to the biosphere will characterise the environment in which we will live in the decades to come. The energy policy of the European Union brought to the fore environmental issues and the issue of decreasing the dependence on imported energy, resulting in the requirement for the maximum possible use of renewable energy sources. Biomass is a very significant renewable energy sources, which is used not only in the energy industry but also in technologies of ferrous metallurgy. The energy potential of biomass can be successfully utilised in the pyrometallurgical processes, which has been partially predicted and confirmed by existing research and studies.

The publication "Biomass and Carbon Fuels in Metallurgy" is mainly oriented on the use of carbonaceous (traditional and alternative) fuels in the metallurgical sector. It describes the application use of these fuels in different technological processes for the production of pig iron, steel and ferroalloys. In this publication, the emphasis is placed on the biomass and its metallurgical utilisation.

The introductory part of the publication "Biomass and Carbon Fuels in Metallurgy" is focused on the specification of fuels, their classification and the characteristics of their basic properties.

Another section is dedicated to the utilisation of traditional carbonaceous fuels in the production of agglomerates, pig iron, steel and ferroalloys. An important part focuses on the use of these carbonaceous fuels in the production of various kinds of agglomerates (ferriferous, manganese and metallised).

A significant part of the publication is oriented on alternative fuels – particularly biomass and its utilisation for metallurgical purposes. It describes characteristics of biomass, techniques of its treatment and the possibilities of its use in the production of charge for blast furnaces, pig iron, steel and ferro-alloys – with emphasis on the use of biomass for the production of ferriferous agglomerates.

The final part of the publication "Biomass and Carbon Fuels in Metallurgy" is focused on defining the relationship between biomass and the environment and the prediction of potential new alternative fuels of the future.

The authors of "Biomass and Carbon Fuels in Metallurgy" have long-term experience in the field of production of various kinds of agglomerates. Recently, they have also been working in the area of biomass use as a substitute for coke powder in the production of the ferriferous agglomerate. Since this area is still rather unexplored and the significant advantages and disadvantages of biomass utilisation have not been fully defined, the authors decided to offer their experience with the use of carbonaceous fuels in metallurgy both to the expert community and the general public.

Biomass represents great potential for the metallurgical industry – although its use and utilisation for specific technological fields of metallurgy are limited. Biomass should be perceived in the context of the relationship between human society, nature, and the production cycle, where its representation should be in accordance with ecological and economic needs of a manufacturing enterprise. By effective use of even a small amount of wood or plant matter from secondary materials in metallurgy, we gain not only the benefit of saving a certain amount of conventional fossil fuels, but we mainly achieve the potential and experience in operating the metallurgical production technologies using alternative fuels. The entire development of metallurgical production technologies using alternative fuels would not be possible without sufficient human resources. It is therefore important to ensure the training of experts for future work in this area. A good theoretical basis is the prerequisite for the education of these professionals. The publication "Biomass and Carbon Fuels in Metallurgy" is thus intended both for lecturers and students of technical universities, researchers, experts from practice, and also for the general public interested in the issue of carbonaceous fuels and their alternatives and implementation in metallurgy. The publication can be used not only in metallurgy but also in the fields of energy, ecology and other technical scientific disciplines.

2. Fossil fuels and renewable energy sources

Energy

Energy is the basis of all processes that take place around us and is the most critical factor influencing the development of society. Energy is a phenomenon without which life is inconceivable today. Without energy, we could not use most of the production facilities and household appliances, set vehicles in motion and we would not have heat and light. If we look at today's world through the prism of the main and priority areas on which we should focus, then there is certainly a question of ensuring low-cost and sustainable forms of energy. Yes, it is necessary to find, transform and utilise these forms of energy, but it is necessary to do it ecologically. So, let us try to extend the phenomenon of energy into the phrase "energy and environment".

How has the humankind changed in the last 100 years? How have production technologies changed during that time? How has the range of manufactured products changed over that period? How has the people's view of the need for energy and its use at their work and in their daily life changed? And how has our environment changed in the context of energy production and subsequent manufacturing activity? How long will the basic fossil fuels last? Will we be able to replace these fuels completely? These are the questions we could answer at length, but also with a simple and short sentence: "We live in dynamic times – with high energy consumption". Those who see the threat of basic energy sources depletion and climate change associated with greenhouse gas emissions know the meaning and the importance of finding new forms of energy. Some traditional energy sources are being gradually consumed and are increasingly expensive. Therefore, we have searched for alternative forms of energy sources for decades. Maybe it is appropriate to ask a question, whether it would not be better to find and develop such technologies and products that would require minimum energy instead of searching for alternative forms of energy sources. Or to implement effective energy saving, which represents the cheapest "source of energy".

Energy sources

Non-renewable and renewable energy sources are used in various technological processes in the production of the required forms of energy. The non-renewable energy sources include mainly fossil fuels based on coal and hydrocarbons. Fossil fuels may be present in solid (e.g. coal), liquid (e.g. petroleum products) or gaseous (e.g. natural gas) form. The non-renewable energy sources also include nuclear energy, *Fig. 1.*

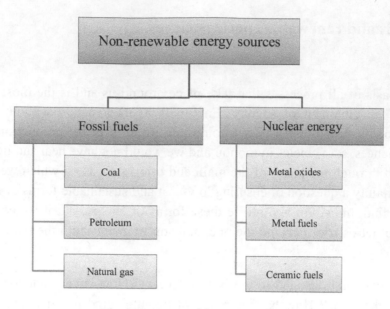

Fig. 1 Non-renewable energy sources

Currently, the most important renewable energy sources are hydropower, biomass and solar energy, Fig. *2* [1–3].

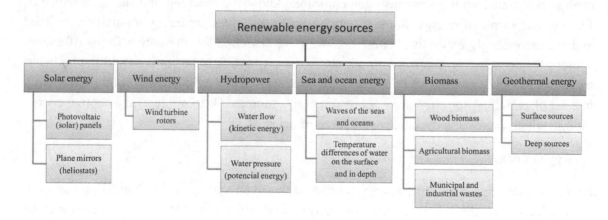

Fig. 2 Renewable energy sources

Fig. 3 shows a graphical view of the energy consumption according to the energy sources used in the world. Apparently, the consumption of traditional fossil fuels was considerably predominant in the last 40 years. Fossil fuels are the predominant source of energy, and they account for total world energy consumption of around 87 % (in 2013). Hydropower has the highest representation among the renewable energy sources. So far, biomass has only a minimum share in the total energy production and consumption.

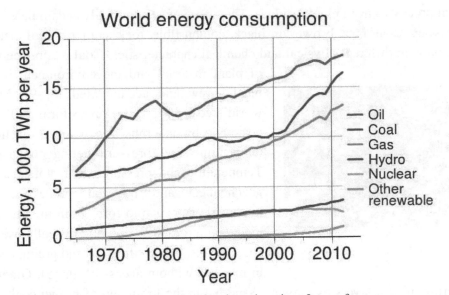

Fig. 3 World energy consumption through various forms of energy

The genesis of fossil fuels

The basic solid fossil fuels include peat, brown coal, lignite, bituminous coal (*Fig. 4*) and anthracite (*Fig. 5*). Coke is the special type of fossil fuel (*Fig. 6*). Coal, coke, petroleum products and natural gas represent the highest quality sources of fossil fuels and are among the most important sources of energy that are currently used.

Fig. 4 Fossil fuel – bituminous coal

In terms of raw material energy, there is a significant potential in so-called unconventional fossil fuels like petroleum and coal shale, biomethane, heavy oil sands, asphaltenes and the like. [4–7].

Let us look in more detail at the processes of the genesis of individual fossil fuels. Peat is formed from plant residues on the surface of Earth, while the coal formation takes place in the depth of Earth. Peat is the predecessor of coal, as it is the first step of converting plant matter into coal. In the long term, peat is primarily formed by biochemical processes, whereas coal is formed by geochemical and physical processes [5]. Peat is divided by prevailing botanical composition into moss and wood, and by soil admixture into pure and natural. Peat has high water content

Fig. 5 Fossil fuel – anthracite

(65 to 85 %). Therefore, it is processed before use by pressing and drying to the water content below 25 %. Coal is a non-renewable fossil fuel, which was created by means of natural

geological processes from plant residues. They were most commonly accumulated in swamps, lakes or seas. Coal is a brown or black combustible rock composed of heterogeneous components with different physical and chemical characteristics. Coal retains the information

Fig. 6 Fossil fuel – coke

of plant material and the environment in which the plants grew and accumulated [8, 9]. Most of the world's coal reserves began to form 300–400 million years ago by anaerobic biochemical reactions at high temperature and pressure acting for a long time. Temperature and pressure are factors that affect the geochemical and physical processes. Coal and hydrocarbons began to form from the original organic material, represented mainly by cellulose ($C_6H_{10}O_5$) and asphalts, by peatification and gradual coalification in marshes without access of air [5]. Gradual descent and covering of peat layers by clay and sand led to the formation of brown coal. Bituminous coal and anthracite have been formed by the action of strong tectonic pressure and high temperature. The whole carbonification process is characterised by diagenesis (a change that results in the conversion of loose materials into a compacted rock) and metamorphosis (rock transformation changing its physical-chemical properties) [5]. The hydrogen and oxygen content decreases and the carbon content increases. The carbon content in the combustible increases from 35 % (content in peat) to more than 92 % (content in anthracite). Coke has even higher carbon content in the combustible – about 94 %.

As mentioned above, individual types of fossil fuels differ by the degree of coalification and chemical composition, *Fig. 7* and *Tab. 1*. *Fig. 7* also presents the factor of coal deposit depth, showing, e.g. that brown coal is located at the depth of 4–6 km, bituminous coal at 6–8 km, and anthracite at up to 10 km.

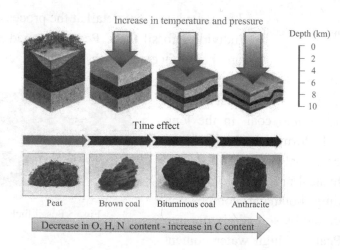

Fig. 7 Impact of factors on the formation of solid fossil fuels [modified by authors according to 10–12]

As already mentioned above, we include coke among fossil fuels. Coke is actually a processed fuel that is industrially produced by the carbonisation of coal, without access of air in so-called coke batteries. Carbonization of coal enables the more economical use of coal by producing better quality fuels and at the same time also raw materials for metallurgical and chemical industries.

Tab. 1 Typical composition of solid fossil fuels

Type of fuel	H_2O [%]	Ash [%]	Volatile matter in combustible [%]	Carbon in combustible [%]	Sulfur [%]
Peat	> 60	1–20	70–90	35–40	0.1–1.0
Lignite	40–50	5–15	50–60	50–60	0.2–1.5
Brown coal	20–40	5–25	40–60	60–75	1.0–3.0
Bituminous coal	2–15	2–15	10–40	70–90	0.1–2.0
Anthracite	1–5	5–12	3–8	87–94	0.3–1.5
Coke	1–3	8–14	1–3	92–97	0.3–0.8

Crude oil as the basis of liquid fuel products is also included among fossil fuels and is composed of liquid hydrocarbons produced by the decomposition of organic material on the bottom of seas. Natural gas is a gaseous form of fossil fuel. It is a mixture of gaseous hydrocarbons (mainly methane – CH_4), enclosed in the ground or escaping from it.

Renewable energy sources

Renewable energy sources are defined as continuously replenished energy sources of various forms. This energy comes directly or indirectly from the Sun or the heat generated deep inside the Earth. Individual renewable energy sources are produced from the Sun, wind, water (watercourses on the land, water of the seas and oceans), biomass and geothermal sources [1].

Solar energy is the basis of almost all renewable energy. This energy has a crucial impact on the weather conditions on Earth. It is inexhaustible and environmentally clean. The Sun as one of the stars in our galaxy is a highly stable and highly powerful energy source. The energy of the Sun originates in the proton-proton fusion reaction. The reaction takes place at temperatures of 14 million degrees Celsius while the surface temperature of the Sun is about 6000 °C. Solar radiation can be used to generate electricity using photovoltaic (solar) panels (*Fig. 8*) and flat mirrors (i.e. heliostats). At present, photovoltaic cells are used the most, which are made of crystalline silicon in the form of monocrystal or polycrystal. Efficiencies of photovoltaic cells are at the level of about 12–20 %. The highest amount of energy from solar panels in the world is produced in Germany and Italy.

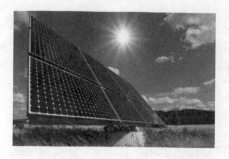

Fig. 8 Photovoltaic (solar) panels

The Sun warms the Earth's surface and atmosphere and therefore layers with different temperatures and pressures are created in the atmosphere. This causes the flow of air, creating wind that can be used as a renewable source of energy in rotors, *Fig. 9*. The highest utilisation of wind energy occurs at airflow speeds around 12 $m.s^{-1}$. Efficiencies of rotors at wind farms are at about 40–45 %. Wind power is used the most in the USA, China and Germany.

The Sun radiation evaporates water of the rivers, seas and oceans. The water vapour condenses in the upper atmosphere and falls back on the surface of Earth in the form of

Fig. 9 Wind turbine rotors

raindrops. It is, therefore, another renewable energy source, where the energy of water flowing in watercourses is used. The energy of water is recovered in hydro power stations not only through its flow (kinetic energy) but also by utilising its pressure (potential energy), *Fig. 10*. The advantage of hydro power stations is their ecological nature and safety. The efficiency of hydroelectric turbines is about 95 %. China, Brazil and Canada produce the highest amount of energy in the world using hydropower.

The gravity of the Moon causes the tide of the oceans and wind creates waves on the seas and oceans. Temperature differences of water on the surface and in the depths of the oceans can also be utilised. All of these forms of energy can be used and represent another source of renewable energy – the energy of the seas and oceans.

Fig. 10 Hydro power station

All plants use carbon dioxide (CO_2) from the atmosphere, and water from the ground for growth. These are converted into organic biomass matter (cellulose ($C_6H_{10}O_5$)$_n$ or glucose $C_6H_{12}O_6$) through photosynthesis. The photosynthetic reaction can be expressed by the chemical equation (1):

$$12\ H_2O + 6\ CO_2 + \text{sunlight} \rightarrow C_6H_{12}O_6 + 6\ O_2 + 6\ H_2O \qquad (1)$$

Solar energy, due to which photosynthesis takes place, is present in chemical bonds of the hydrocarbon material that is called biomass. By burning biomass, we obtain energy, while carbon dioxide (CO_2) and water are formed. The process is cyclic, continuous and renewable because resultant carbon dioxide (CO_2) is the input component for the growth of a new plant, i.e. new biomass. Although the combustion of biomass releases environmentally unacceptable carbon dioxide into the atmosphere, it is only the quantity that has been accumulated by photosynthesis in plants. Therefore, we say that biomass combustion is "CO_2 neutral", *Fig. 11*. In relation to the zero balance of CO_2 in the production and management of biomass, it is necessary to note that it is only a comparative term. In fact, taking care of growing biomass, its subsequent treatment, transport and processing are activities releasing carbon dioxide. A neutral balance of CO_2 in connection with the processing of biomass is therefore never achieved. However, the contribution of biomass processing to the overall balance of CO_2 in the energy production compared to fossil fuels is incomparably smaller.

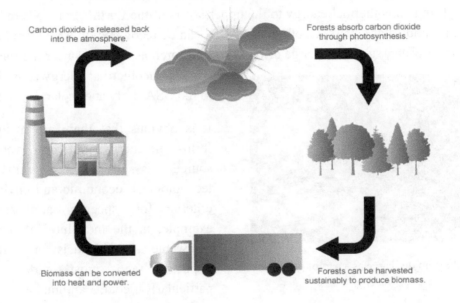

Carbon dioxide is released back into the atmosphere.

Forests absorb carbon dioxide through photosynthesis.

Biomass can be converted into heat and power.

Forests can be harvested sustainably to produce biomass.

Fig. 11 Cycle of carbon dioxide (CO_2) in nature

In this context, it is perhaps appropriate to ask why carbon dioxide is perceived as harmful to the environment. Everything that happens in nature should be in balance. Also, the amount of CO_2 that enters the atmosphere through various processes (including production processes) should be consistent with the absorption ability of other chemical processes. Besides carbon dioxide and water, plants need also oxygen and other biogenic elements (e.g. phosphorus, potassium, calcium, magnesium, sulfur and iron) for life in the form of minerals. If the amount CO_2 in the atmosphere is above the limit, the amount of vital oxygen in the air decreases. Thereby it disturbs the balance in nature and the ability of plants to process these excess amounts of carbon dioxide through photosynthesis.

Fig. 12 Biomass – wood products

In this publication, we will focus on biomass and its use much more. First, it is necessary to define what biomass is. Biomass is a material of organic origin, which includes wood (*Fig. 12*) and plant biomass grown in soil and water, livestock biomass and organic waste [1, 13]. The USA, Germany and Brazil produce the highest amount of energy in the world using biomass.

Although geothermal energy is not really the renewable source, it is classified as a renewable source, because of its inexhaustible supply. Geothermal energy originates in the hot Earth's core, the temperature of which is estimated at about 4000 °C. A large amount of thermal energy is released by radioactive decay of substances and gravitation forces. The movement of magma brings this thermal energy to the upper layers of the Earth's crust where heating of rocks and groundwater occurs. Geothermal energy can be recovered directly from hot springs of geothermal water, steam discharge or by means of boreholes in rocks near the Earth's crust

Fig. 13 Steam discharge as a part of geothermal energy

Fig. 13. Geothermal energy is used the most in the USA, Italy and Iceland.

It is obvious that there are many various forms of energy, but not every energy source is consistent with current technological, economic and environmental criteria for specific application. For example, in the metallurgical sector, non-renewable fossil fuels are the most important and most widely used – particularly coal, anthracite, coke, oil products and natural gas.

At the beginning of this chapter, the concept of energy and forms of energy were explained. But what is the fuel? We will come across this concept quite often in the following chapters of this publication. Fuel is the common name for a chemical element, chemical substance or a mixture of substances, which can react in chemical reactions of combustion under certain (precisely defined) conditions. In these reactions, the chemical energy contained in fuel is released and transformed into thermal energy.

3. The role of fuels in metallurgy

Metallurgy and metals

Energy sources and fuels are used mainly in the energy sector for the production of heat and electricity, but they also play an important role in metallurgy. Metallurgy is one of the oldest industrial and scientific fields, and its primary concern is the production of utility metals and alloys. Production of metals is inconceivable without energy and fuels. Before we can define the need for fuel in metallurgy, let us introduce the metallurgical world. It is very comprehensive and interesting, and metals have original and various properties. Their extraction is based on pyrometallurgical or hydrometallurgical processes. For example, magnesium is a very light and highly reactive metal. It easily captures oxygen, sulfur and chlorine from most of the alloys. In addition, magnesium alloys are used for the production of cars and aeroplanes. Aluminium is one of the most significant materials in the automotive and aircraft industry. Titanium is characterised by very high hardness and strength. Titanium, being two times lighter than iron, is stronger than many types of steel. Chromium is the hardest metal, but it is also very resistant to atmospheric influences. Stainless steel, which is highly corrosion resistant, contains about 17–19 % of chromium. Manganese also has high hardness, high resistance to wear, and so-called Hadfield steel (with the manganese content of 12 %) becomes even tougher by mechanical impacts. Radioactive cobalt is used, for example, in non-destructive defectoscopy, i.e. in quality control of various products. Copper and silver have high electrical and thermal conductivity and are therefore often used in electrical engineering. Gold is chemically very stable and does not change even in strong acids. Platinum is used in the production of thermocouples, whereas the electrical resistance of platinum increases linearly with temperature. Tungsten is a metal with the highest melting point (3400 °C) and is used mainly in electrical engineering.

Black metallurgy, i.e. metallurgy focused on the production of iron, steel and ferroalloys, is one of the essential fields of metallurgy. Iron has been one of the most significant metals throughout the entire history of the universe. Without iron, there would be no life on Earth. Moreover, the alloy of iron with carbon and other additives (steel) is the most important construction material. Iron is an essential component of haemoglobin (a substance that supplies the living fibres with oxygen) and can be found in all plants, where it is necessary for the production of chlorophyll (chlorophyll plays a vital role in converting light energy into chemical energy in the photochemical process in plants – photosynthesis). Iron and its alloys are used in many applications and many products (bridges, rails, equipment, tools, cars, etc.). Iron currently remains the basis of metallurgy, engineering, civil engineering and transport. Today, around 2000 types of steel are produced worldwide for variety of purposes – e.g. stainless steel, high-speed steel, ball bearing steel, spring steel, magnetic and non-magnetic steel, creep resistant steel, low-temperature steel or diamond steel (contains about 5 % of tungsten and its hardness is only slightly lower than that of the diamond). There are so many types of steel that their enumeration would require more than ten pages of this publication...

Metallurgy and fuels

Whereas manufacturing of iron-based products is included among pyrometallurgical technologies, it is necessary to use heat sources and fuels in these the metallurgical processes. Pyrometallurgical technologies are high-temperature manufacturing techniques that require heat. The heat can be obtained by conversion of various forms of energy, e.g. electrical energy, energy of carbon from hydrocarbon fuel combustion (chemical energy) or by the regeneration and recovery of waste heat, *Fig. 14*.

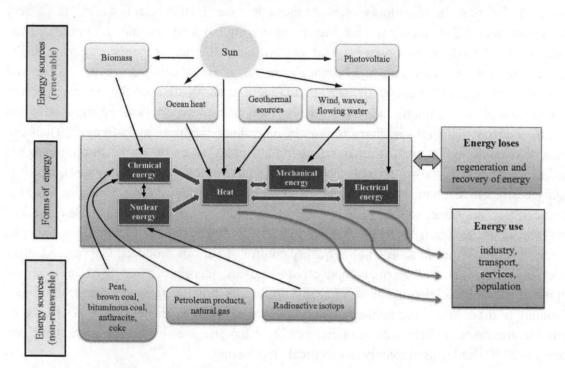

Fig. 14 Energy sources and forms of energy conversion [modified by authors according to 14, 15]

This publication focuses mainly on the use of carbonaceous fuels. In black metallurgy, these fuels may be used as a heating medium, carburiser, and reducing agent. In addition to these three primary uses of carbonaceous fuels, they are also used for manufacturing of electrodes (söderberg self-baking, carbon and graphite) for the EAF. *Fig. 15* shows basic forms of use of carbonaceous fuels in individual technologies of black metallurgy. Individual carbonaceous fuels should have optimal properties for specific technologies. These will be discussed in the following chapters of this publication.

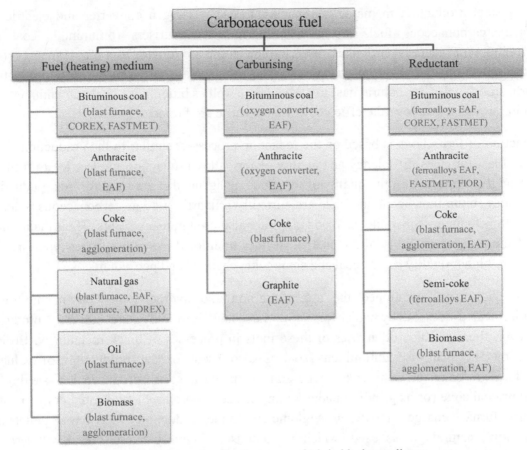

Fig. 15 Various uses of carbonaceous fuels in black metallurgy

(Legend: In the technical and technological practice, a fuel represents not only heating medium, but can also be used for other purposes, for example carburising, reduction, production of electrodes, etc.)

Production of iron, steel and ferro-alloys takes place in furnace reactors, for which it is necessary to provide a variety of carbonaceous fuels. Almost any of the non-renewable fossil fuels (or products of their refining), as well as, e.g. renewable biomass, may be used as carbonaceous fuel. Metallurgical coke is required for the production of pig iron in a blast furnace (BF), but other carbonaceous fuels, e.g. bituminous coal and anthracite, are also used. In addition to carbonaceous fuels, hydrocarbon fuels, such as natural gas and oil, are also used in blast furnaces.

All these fuels should ensure a sufficient amount of heat in the blast furnace, which is produced by their combustion. Blast furnace coke is the most important carbonaceous fuel in the blast furnaces. However, it also serves another purpose besides its calorific value. Since pig iron is produced by reduction of ferriferous materials (iron ore, sinter and pellets), coke is also used as a reducing agent in a blast furnace. Furthermore, the metallurgical coke contributes to the carburization of iron and forms the so-called framework (stability) of a charge in the blast furnace as well. It is obvious that metallurgical coke – as an essential carbonaceous fuel in the blast furnace – plays its irreplaceable role in this type of furnace.

During steel production in major steelmaking furnaces (oxygen converter and electric arc furnace), carbonaceous fuels are used primarily as carburisers (bituminous coal and anthracite). Anthracite is particularly used in the electric arc furnace as a reducing agent (e.g. for the reduction of FeO from steelmaking slag) to increase the yield of steel production. Hydrocarbon fuels (e.g. natural gas) in particular as well as bituminous coal and anthracite are used in burners to improve the efficiency of the electric arc furnace.

Production of ferroalloys is based on the reduction processes, similar to the production of pig iron in blast furnaces. Efficiency and quality of ferroalloy production in electric arc furnaces are strongly dependent on the quality of the reducing agents that are used for this production. In practice, bituminous coal and coke are currently primarily used as carbonaceous reducing agents for producing ferro-alloys. Wood chips are also used in many plants. Their main role is to enhance the permeability of the charge. Besides traditional reductants, other types are also used, e.g. higher quality brown coal and semi-coke, charcoal and petroleum coke.

The previous text mentioned the use of carbon and hydrocarbon fuels in the major metallurgical processes, by which pig iron, steel and ferro-alloys are currently produced. However, there are also other uses of these fuels in processes of black metallurgy. Besides direct metallurgical use, bituminous coal is also used for the production of valuable metallurgical coke or semi-coke. Fine-grained powder coke, produced by sieving of metallurgical coke (or its grinding under 3 mm), is used to produce an essential component of the blast furnace charge – ferriferous agglomerate. In the production of ferriferous pellets in a rotary kiln, natural gas is used, which is combusted in the burners. Globally, there are alternative metallurgical technologies (e.g. COREX, HISMELT, MIDREX, FASTMET, FIOR, ITmk3, etc.) that are utilised to produce products (e.g. liquid iron, DRI, HBI) with a very similar chemical composition to pig iron. These technologies also use carbonaceous and hydrocarbon fuels – e.g. bituminous coal, anthracite and natural gas. Graphite is also used in the production of casting powders for the steel casting technology.

In general, carbon is the basic solid carbonaceous fuel in black metallurgy. Its use in current technologies is graphically documented by the scheme in *Fig. 16*. The processes of direct iron production from ores (e.g. FASTMET, FASTMELT and ITmk3), preheating of scrap in EAF, carburising of steel, production of electrode paste, production of carbon composite electrodes and a reducing agent in the production of ferroalloys in the EAF can be included among the progressive technologies, which will use high quality coal (bituminous coal and anthracite) on a larger scale in future.

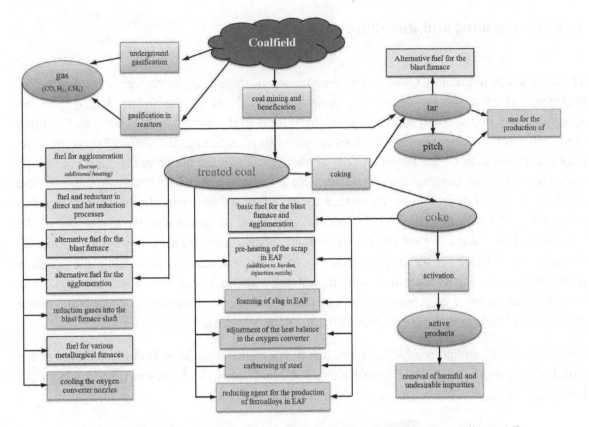

Fig. 16 Utilization and use of coal in metallurgy [modified by authors according to 16]

All of the mentioned metallurgical technologies are inconceivable without carbonaceous fuels. However, the gradual replacement of non-renewable fossil fuels with renewable fuels – particularly biomass, which is composed of carbon and hydrocarbons – is conceivable, and it is already taking place. Biomass consists exactly of the valuable material that we need to supply to the iron, steel and ferroalloy-making processes.

4. CO_2 emissions and metallurgical technologies

The question of reducing CO_2 emissions into the atmosphere is currently very urgent and very important. Human society causes the emissions of carbon in the form of CO_2 into the atmosphere by mining, burning and use of fossil fuels in the metallurgical cycle, and disturbs the equilibrium concentration of greenhouse gases in it. Although the share of these gases is only one-thousandth of the volume of the atmosphere, the greenhouse gases (carbon dioxide, methane, nitrogen oxides, water vapour, etc.), which are natural constituents of the atmosphere, are of an immense importance for us. Their concentration in the atmosphere was balanced by the natural carbon cycle and thus prevented its accumulation in any part of the biosphere. The release of ever-increasing amounts of greenhouse gases, however, continually disrupts the balance between the incoming and emitted energy, increasing the average temperature on Earth (although existing studies do not confirm this directly). The metallurgical industry as a part of the industrial sector is the third largest contributor to the production of carbon dioxide emissions after the energy and transport sectors [17].

Fig.17 shows the annual production of carbon dioxide emissions in individual countries of the world. China produced up to 27 % of global CO_2 emissions (18 % in the USA, 13 % in the European Union, 0.12 % in Slovakia).

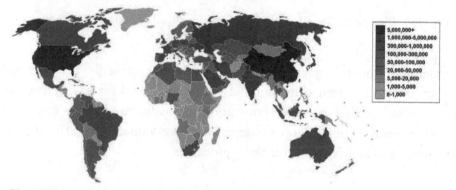

Fig. 17 CO_2 emissions in individual countries in 2012 (thousands of tonnes/year) [18]

It is necessary to recognise that the combustion of 1 kg of bituminous coal produces 2.56 kg of CO_2. By burning of 1 kg of diesel, 3.12 kg of CO_2 are released, and the combustion of 1 m^3 of natural gas results in 2.75 kg of CO_2. These values are irreversible; hence several billion tonnes of CO_2 emissions are produced in the world each year. The process of reducing CO_2 emissions (or at least slowing down their growth related to the constant increase of industrial production) is now relevant also for primary metallurgical production of iron and steel. Nonetheless, this issue is problematic, as it is currently not addressed globally for the entire sector of metallurgy in the world, but only for certain regions. The consequence of this state would be the relocation of metallurgical steelmaking (e.g. from Europe and the USA) to countries where CO_2 emissions are not an issue yet (e.g. India, China, South American countries). In the USA and European Union countries, 1.3 to 1.8 tonnes of CO_2 are produced per tonne of steel. In India and China, it is almost twice as much. Although CO_2 emissions have been significantly reduced in the metallurgical sector of some countries (at most by 40 %) over the last 30 years, the continuous increase in the volume of steel production on a global scale will force the sector into yet more significant reduction of CO_2. This can be achieved through the research and development of new technologies or restructuring the energy and raw material base of current technologies [17]. The topic of this publication falls within this framework as well.

When comparing the production of steel in an integrated metallurgical cycle of a blast furnace – oxygen converter with the cycle of mini-mill smelting based on an electric arc furnace, the integrated cycle seems to be the more energy-intense and ecologically negative. The energy intensity of steel production in the integrated metallurgical cycle represents approximately 10.5 to 11.5 GJ/t steel (in certain efficient and restructured facilities in the EU, USA and Japan) compared to 14–18 GJ/t steel (in some facilities in Southeast Asia and South America). The theoretical minimum amount of energy required to produce 1 tonne of steel was calculated to be 7.9 GJ. The quantity of CO_2 per 1 tonne of steel in the integrated metallurgical cycle is about 1350–1750 kg (in certain facilities of Southeast Asia and South America, this value is considerably higher). The theoretical minimum amount of CO_2 in the integrated cycle was calculated to be 1230 kg/t steel. It is clear that the intensity of steel production in the integrated metallurgical cycle is directly related to CO_2 emissions. The amounts of CO_2 at the level of theoretical minimum values can be achieved in the integrated metallurgical cycle of iron and steel production by recycling of energy sources and the use of secondary raw materials and biomass. Increasing the energy efficiency of technology in integrated metallurgical enterprises results in a reduction of emissions (CO_2, NO_X, SO_X).

Tab. 2 shows the CO_2 emission factors in individual operations of primary iron and steel production. These emission factors correspond to the average value of some EU plants in the years 2008–2012. They differ from case to case. It is known that pig iron production in blast furnaces contributes to the CO_2 production the most, as it consumes the largest amount of fossil fuels as well (metallurgical coke, bituminous coal, anthracite and natural gas).

Tab. 2 CO_2 emission factors for the processes of primary iron and steel production [18, 19]

Process [measuring unit]	Emission coefficient CO_2	Source
Coke production in coke oven battery [tonnes CO_2 per tonne of produced coke]	0.56–0.86	European IPPC
Production of agglomerate on sintering strand [tonnes CO_2 per tonne of produced agglomerate]	0.20–0.35	European IPPC
Production of pellets in rotary kiln [tonnes CO_2 per tonne of produced pellets]	0.02–0.03	European IPPC
Production of pig iron in blast furnace [tonnes CO_2 per tonne of produced pig iron]	1.15–1.35	European IPPC
Production of steel – BF/BOF cycle [tonnes CO_2 per tonne of produced steel]	1.36–1.50	International Iron and Steel Institute
Production of steel – EAF cycle [tonnes CO_2 per tonne of produced steel]	0.10–0.30	International Iron and Steel Institute

The process of reducing CO$_2$ emissions from the metallurgical technologies is now actively pursued only in Europe, North America, Southwest Asia and Japan. The ULCOS project (reduction of CO$_2$ emissions in the iron and steel metallurgy by 50 % by 2015) provides for the use of biomass, replacing carbon-based reductants with hydrogen from electrolysis of water, and capturing (storing CO$_2$) [20]. The assessment of feasible projects suggests that technologies based on recycling of energy sources and secondary raw materials and the replacement of coke with coal and natural gas will be the primary means of greater reductions in CO$_2$ emissions in the integrated iron and steel cycle in the near future, Tab. *3*. The new technologies based on non-carbonaceous reagents will also be interesting (their development and impact on iron and steel technologies are difficult to predict today).

Tab. 3 Energy saving and CO$_2$ emissions reduction by the intensification of the processes of primary iron and steel production [21]

Intensification	Energy savings [GJ/t steel]	Decrease of CO$_2$ emissions [kg C/t product]
Agglomeration		
Recuperation of flue gases heat	0.12	3.41
Increasing the thickness of sintered layer	0.02	0.59
Increase of process control – automation	0.01	0.3
Use of waste gases and secondary raw materials (sludge and dust from BF, BOF)	0.08	2.16
Partial substitution of coke powder by biomass	research	research
Coke oven		
Precise coal moisture control	0.09	0.55
Programmable chambers heating	0.05	0.31
Dry coke quenching	0.37	2.25
Blast furnace		
Pulverised coal injection up to 130 kg/t p.i.	0.69	11.42
Injection of natural gas up to 140 kg/t p.i.	0.80	13.35
Automation of wind heaters	0.33	5.49
Increase of process control – automation	0.36	5.99
Partial substitution of metallurgical coke powder by biomass	research	research
Steelworks – BOF		
Recuperation of flue gases heat	0.92	12.55
Injection of inert gases	0.22	3.20

Another possibility is the cross-over from integrated metallurgical cycle to mini-mill smelters that are based on steel scrap recycling (or DRI and HBI) in an electric arc furnace. In case of the intensification and restructuring proceedings in an integrated metallurgical cycle of iron and steel production, a significant reduction of CO$_2$ emissions cannot be expected (max. 30 % because of the need for large amounts of metallurgical coke in this cycle). The development of new technologies (e.g. DRI, HBI based on the reduction by natural gas and coal) and the transition to EAF steel production can achieve more substantial reduction of CO$_2$. As a part of the research, currently preferred source of fuel is wood and plant biomass, which is an attractive alternative of the fuel substitution for industrial use, *Fig. 18*. In research projects, bio-oil and biogas are also being tested. Using biomass may enable the reduction of CO$_2$ in the metallurgical industry (more about this in the following chapters).

a)   *b)*   *c)*

d)   *e)*   *f)*

Fig. 18 Potential sources of biomass for metallurgical use
a) charcoal, b) hemp, c) bio-oil, d) sawdust, e) nutshells, f) biogas

Besides CO$_2$ emissions reduction in the iron and steel production, this process takes place also in the production of ferroalloys in the EAF. In many factories, part of coke is replaced with biomass products – e.g. charcoal and wood sawdust. The issue of reducing CO$_2$ emissions will be present and discussed in many other chapters of this publication. In this chapter, we have endeavoured at least to introduce this vital subject to the reader, pointing out the priority areas of the world today (including the metallurgical one).

5. Distribution of fuels and their reserves

In the third chapter "The role of fuel in metallurgy", some of the carbon and hydrocarbon fuels currently used in black metallurgy as fuel (heating) media, carburising and reducing agents were specified. In terms of their formation, fuels are divided into renewable and non-renewable. With regard to their state, carbon and hydrocarbon fuels can be classified as solid, liquid and gaseous. Fuels are divided by the origin into natural fuels (directly usable) and modified (thermally or chemically processed). Pursuant to the many years of established technological practice, we can also divide fuels into traditional and alternative. General classification of fuels is shown in *Fig. 19*.

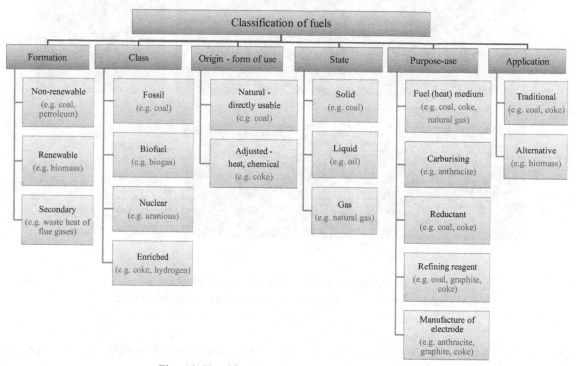

Fig. 19 Classification of fuels by various criteria

The basic classification of fossil fuels is by their state, *Fig. 20*. Most of these fuels are presently usable in the metallurgy of iron, steel and ferro-alloys (e.g. various kinds of coal, coke, oil, natural gas, coal gasification gas and reducing gas from natural gas).

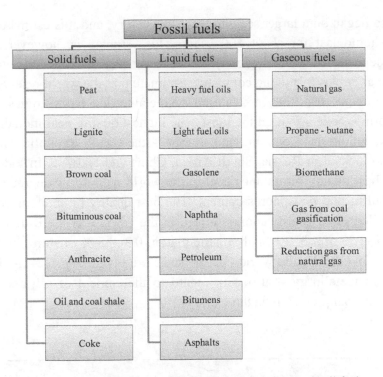

Fig. 20 Classification of non–renewable energy sources – fossil fuels

Fossil fuels are included among the non-renewable energy sources, and the most significant ones are coal, oil and natural gas. These are still indispensable and most widely used sources of energy, and products from these fossil fuels are used to the greatest extent in metallurgy as well.

Coal

The fossil fuel group includes especially coal, which is a mixture of organic and inorganic components. The formation of coal was influenced by number of different factors – e.g. original organic material, time conditions, physical-chemical, biochemical, geological and other conditions. The sum of these conditions affects the variability of physical, chemical and technological properties of coal. Therefore, coal is very inhomogeneous, and its characteristics vary even within a single deposit. Coal is mainly composed of organic compounds called lithotypes (vitrain, durain, clarain and fusain) or microlithotypes (maceral groups, vitrinite, liptinite and inertinite) and a certain amount of minerals [5]. Most of the world coal reserves began to form 300–400 million years ago by anaerobic biochemical reactions at high temperature and pressure. The gradual transformation of plant material into coal occurred according to the scheme: peat – lignite – brown (sub-bituminous) coal – black (bituminous) coal – anthracite [5, 7].

The coal mining began on a larger scale in the 17th century, and it is estimated that the world coal reserves should last for about 500 years. This estimate is, however, very optimistic because many deposits are technologically and economically unextractable even in the longer term. Professional circles of mining companies estimate that the world reserves of extractable coal will last for about 100–200 years. Within the metallurgical technologies, bituminous coal or its refined form – coke – is used the most. The highest quality coal includes coal mined in Australia, USA, Indonesia, Colombia and Europe – the Czech Republic and Poland. Large deposits are also located in Russia, South Africa, China and Ukraine. In Indonesia, there are deposits of the best quality bituminous coal in the world with a low degree coalification and very low ash, sulfur and phosphorus content. The best quality bituminous coal in the world with a medium degree of coalification is mined in the USA. It is the Blue Gem coal, which contains about 1–3 % of ash, 30 % of volatile matter, 0.3–0.5 % of sulfur and about 0.002 % of phosphorous. In *Fig. 21*, the world coal deposits are given. *Fig. 22–23* show the highest quality (usable in metallurgy – also with regard to the content of ash and sulfur) and the largest bituminous coal producers in the world [6].

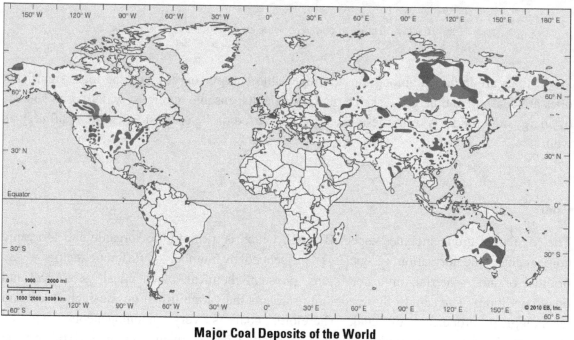

Major Coal Deposits of the World

Anthracite and Bituminous Coal   Lignite

Fig. 21 Illustration of the world coal deposits [6]

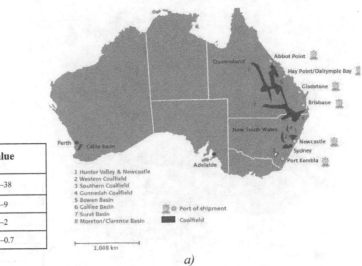

Property of coal (Australia)		Value
Content of volatile combustible	[%]	21–38
Content of ash	[%]	4–9
Content of moisture	[%]	1–2
Content of sulfur	[%]	0.3–0.7

a)

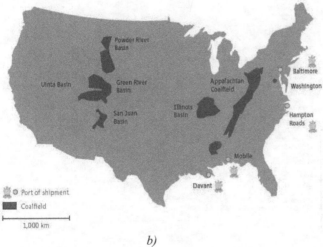

Property of coal (USA)		Value
Content of volatile combustible	[%]	18–34
Content of ash	[%]	5–9
Content of moisture	[%]	1–3
Content of sulfur	[%]	0.5–1.2

b)

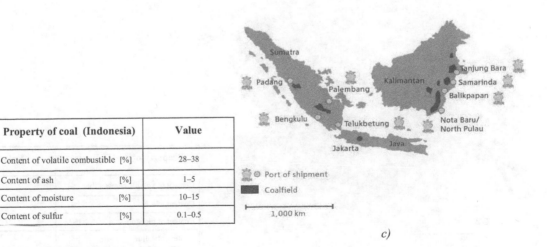

Property of coal (Indonesia)		Value
Content of volatile combustible	[%]	28–38
Content of ash	[%]	1–5
Content of moisture	[%]	10–15
Content of sulfur	[%]	0.1–0.5

c)

Fig. 22 The best quality bituminous coal reserves in the world – a) Australia b) USA c) Indonesia
[22, 23]

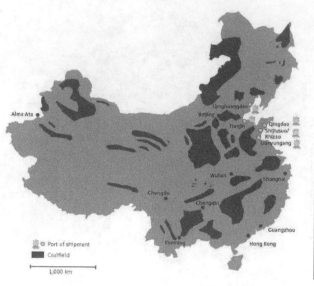

Property of coal (China)	Value
Content of volatile combustible [%]	25–30
Content of ash [%]	8–12
Content of moisture [%]	1–2
Content of sulfur [%]	0.3–0.8

a)

Property of coal (South Africa)	Value
Content of volatile combustible [%]	16–29
Content of ash [%]	5–15
Content of moisture [%]	–
Content of sulfur [%]	0.2–0.8

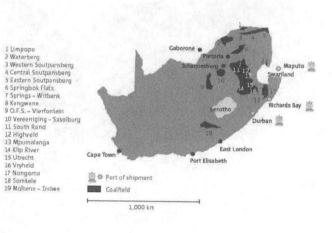

b)

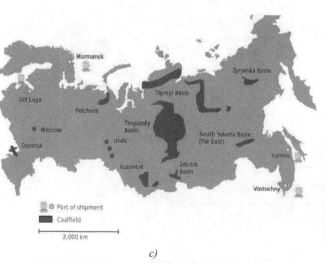

Property of coal (Russia)	Value
Content of volatile combustible [%]	15–39
Content of ash [%]	3–12
Content of moisture [%]	0.1–12
Content of sulfur [%]	0.3–1.2

c)

Fig. 23 The largest reserves of bituminous coal in the world – a) China b) South Africa c) Russia
[22, 23]

The main disadvantage of coal is that by its use in coal-fired power stations or metallurgical processes, a large amount of harmful substances is released into the atmosphere. These are particularly the oxides of carbon, sulfur and nitrogen. Today, technologies that allow producing cleaner and better fuels from coal are coming to the forefront. In addition to environmentally demanding coke production from mixtures of different types of coal, coal is also gasified or condensed. Coal can be condensed on the basis of organic compounds extraction with the aid of solvents and further saturation by hydrogen. This creates a number of products that can be used in metallurgy, e.g. heavy oils, asphaltenes, etc. Coal gasification is carried out underground, directly in the coal bed, or in gasification reactors. In gasification systems with combined cycle (IGCC – Integrated Gasification Combined Cycle), coal is not combusted directly but reacts with oxygen and steam, thus forming a synthetic gas – "syngas". The synthetic gas is combusted in a gas turbine after purification for power generation and produces steam to drive a steam turbine. Produced synthetic gas also has a high reducing capacity and is tested for use in metallurgy – e.g. within reduction processes. Use of the most coalified coal – anthracite – is interesting in the context of metallurgical technologies as well. Its use as a reducing and carburising agent has already been mentioned. Nonetheless, a higher quality product can also be produced from anthracite. So-called calcined (carbonised) anthracite is acquired by calcination (commercial term, the technical term is carbonisation) of anthracite coal. It is used, e.g., in electrode materials for ferroalloy production in the EAF. In this publication, many examples of the various use of different types of coal in the iron and steel metallurgy have been already mentioned. Nevertheless, its main use is for energy purposes. The share of coal in the world energy production is about 25 %.

Crude Oil

It is one of the most important fuels at present. The share of crude oil derivates in the world energy production is about 30 %. Crude oil is composed of liquid hydrocarbons produced by the decomposition of organic material on the bottom of seas 500 million years ago. Crude oil consists of 80–87 % carbon, 10–14 % hydrogen, 0.2 to 3 % nitrogen, 0.05 to 1.5 % oxygen, and 0.05 to 6 % sulfur. The formation of crude oil is accompanied by the formation of natural gas, which is almost always located above the crude oil deposit. It is estimated that at today's rate of crude oil extraction, average global stocks will last for about 40 to 60 years. Crude oil reserves are geographically unevenly distributed. The largest reserves are in politically or economically unstable regions.

Motor fuels, lubricants, heating oil and petrochemical products (e.g. plastics) are produced from crude oil. *Fig. 24* shows the largest oil reserves in the world. Furthermore, the number of currently predicted years of extraction is also provided for individual countries. Since the estimates are based on the current rate of extraction, given the higher crude oil consumption (especially in China and India), it is possible that the reserves will be exhausted before the time declared by mining companies at present.

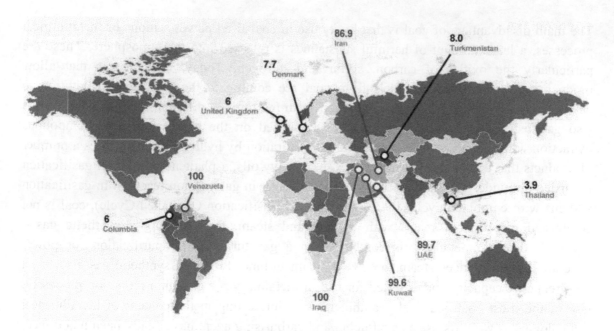

Fig. 24 The largest crude oil reserves in the world [4]
(individual countries with predicted reserves in years)

Natural gas

Natural gas is flammable natural gas used as the primary gaseous fossil fuel. In a broader sense, this term denotes any mixture of gaseous hydrocarbons enclosed under the surface of the Earth or escaping from it, e.g. hydrocarbon gases, CO_2, H_2S and other. In a narrower sense, this term describes the mixture of hydrocarbons, out of which 50 to 90 % is methane by volume. The other components of natural gas are ethane (5–9 %), propane (3–18 %) and the heavier hydrocarbons (2–14 %). Natural gas requires the least adjustments, prior to use, of all fossil carbonaceous fuels. It is cleaned and dehumidified at the mining site and then transported by long-distance pipelines to places of consumption. The share of natural gas in the world energy production is about 27 %.

Fig. 25 shows the largest natural gas reserves in the world, with years of currently predicted extraction for each region. These reserves are predicted according to the current rate of extraction. At significantly higher gas consumption (especially in China), it can be assumed that the reserves will last shorter than declared by mining companies at present. It is estimated that at today's rate of extraction of natural gas, average reserves will be sufficient for about 60–70 years. The situation in natural gas reserves copies the situation in extraction and reserves of crude oil.

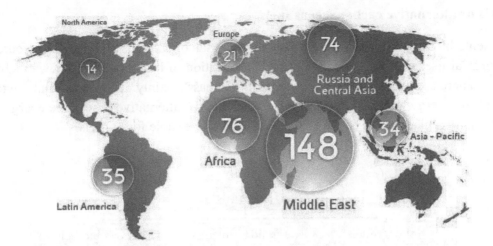

Fig. 25 The largest natural gas reserves (individual countries with predicted reserves in years) [4]

Traditional and alternative carbonaceous fuels

The difference between states of fuels is evident to everyone. The difference between directly usable fuels (e.g. wood, coal) and thermally–chemically modified fuel (charcoal, coke) was explained by the example of charcoal and coke. In the second chapter "Fossil fuels and renewable energy sources", the difference between renewable and non-renewable energy sources and fuels was explained as well. What is then the fundamental difference between traditional and alternative carbonaceous fuels? Which fuels should be included among the traditional carbonaceous fuels and which among the alternative carbonaceous fuel? Before we answer these questions, let us first explain the terms "traditional and alternative". The blast furnace is one of the so-called traditional (conventional or basic) devices, where the pig iron is produced. In addition to this device (which has been the most represented for many decades and by which about 90 % of pig iron in the world is produced), there are also other technological devices, which produce products with very similar chemical composition to pig iron. These technological devices and, consequently, respective technological processes are labelled as alternative. While blast furnace coke is the traditional (conventional or basic) fuel in the blast furnace, it is bituminous coal or natural gas in alternative technology. Coal or natural gas is therefore considered an alternative fuel to blast furnace coke. Both types of fuel have properties that can be used in the production of so-called alternative iron-based products (e.g. DRI, HBI, reduction smelting products, etc.). Similar considerations can also be used in the classification of fuels as traditional and alternative within any technological process. If we gradually replace part of non-renewable fossil fuels (e.g. coal, coke) with renewable fuels (e.g. biomass), renewable fuels will be labelled as alternative fuels. Alternative fuels to fossil fuels can also be called biofuels.

Biofuels or alternative carbonaceous fuels

At present, biomass has the highest usable potential of all renewable energy sources for metallurgical use. *Fig. 26* shows the basic classification of biofuels from renewable biomass by state. There are expectations associated with biomass (mainly dendromass and phytomass) intended for energy and metallurgical use to become the alternative renewable energy source and gradually replace a part of the conventional non-renewable fossil fuels.

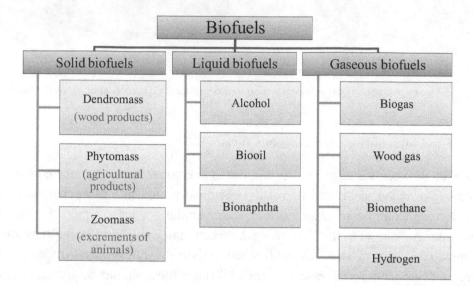

Fig. 26 Basic distribution of biofuels

In terms of energy conversion efficiency, biomass is not only suitable for metallurgical use but also for combustion, production of biofuels in the form of methyl esters of vegetable oils and bio-alcohol as a component of motor fuels, and the production of biogas followed by combined heat and power cogeneration.

When we look at solid alternative fuels from a different perspective, we learn that this category also includes various types of waste in addition to dendromass and phytomass, *Fig. 27–28*. Certain types of waste are based on fossil fuels (e.g. plastics and rubber, which are produced from crude oil), yet they can be classified as alternative fuels.

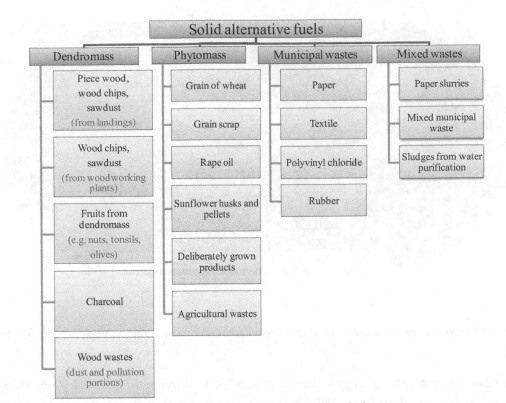

Fig. 27 Distribution of solid alternative fuels

Fig. 28 Phytomass and dendromass (including wood wastes)

The reserves of biomass are currently several times higher than their use for energy purposes and industrial use. Nevertheless, it is necessary to approach the use of these reserves sensitively and environmentally responsibly. Currently used energy of biomass is at the level of about 50 EJ (50.10^{18} J). The technical potential of biomass is at the level of 500–600 EJ, while the theoretical potential of biomass amounts approximately to 2900 EJ. *Fig. 29* shows the overall reserves of biomass in the world, representing about 13 Gha.

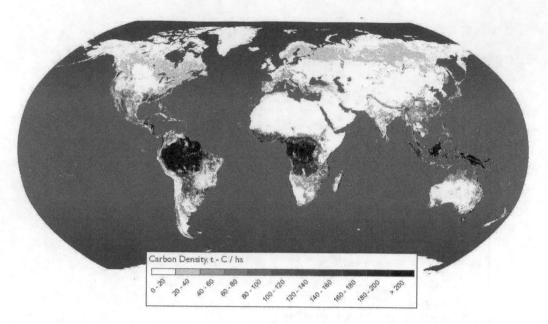

Fig. 29 Reserves of biomass in the world (the amount is converted to carbon equivalent/ha of land) [14]

As biomass is a renewable energy source, it is possible to provide the necessary quantities and properties of fuel for the future. The use of biomass in the metallurgical processes requires consideration of not only securing the needed amounts but especially appropriate selection and modification of the physicochemical properties of the given biomass for a specific use. Current knowledge of biomass potential in the metallurgical industry is rather at the beginning of learning or research. Therefore, the topic of complete replacement of traditional fuels with biomass is a question for the future. Naturally, biomass as a complete replacement (or absolute alternative) of coke is currently not possible. Therefore, in relation to alternative carbonaceous fuel, we are talking rather about partial replacement of traditional (conventional or basic) carbonaceous fuels. The fact that solely a partial replacement of traditional fuels is feasible is related not only to their physicochemical properties but primarily to the physicochemical properties and thermochemical properties of alternative carbonaceous fuels.

6. Carbonaceous fuels, their properties and testing

Precise and substantial evaluation of properties of any matter increases the effectivity of its use and processing. The matter is the objective reality existing independently on our sense, which can be weighted and measured. In the narrower physical meaning, matter is a certain form of substance, which is called the material. All substances consist of atoms, ions and molecules. Each substance has its own qualitative and quantitative parameters. Carbonaceous fuels also consist of basic structural particles of matter and may be differentiated by their properties. The most effective use of carbonaceous fuels as an energy source or raw materials for metallurgical (or chemical) industry requires a substantial evaluation of the properties of this matter.

Solid carbonaceous fuels differ in origin, and their physicochemical properties are closely connected with the formation of materials of these fuels. Basic atoms (elements) of matter of solid carbonaceous fuels are carbon, hydrogen and oxygen (the basic atoms also include nitrogen, but its content in the solid carbonaceous fuel is very low). The content of hydrogen and oxygen in non-renewable solid carbonaceous fuels decreases with increasing degree of carbonisation. On the other hand, in these types of fuel, the carbon content rises in the direction: peat – lignite – bituminous coal – anthracite. The highest quality solid fossil fuels (bituminous coal and anthracite) contain the highest amounts of carbon. Plant and woody biomass (which is included in renewable energy sources) contains the highest amount of hydrogen and particularly oxygen of all existing solid carbonaceous fuels, *Fig. 30*. The calorific value (heating value) of carbonaceous material grows in the direction of the decrease in the H/C and also O/C ratios [11]. The increase in heating value is thus directly proportional to the increase in carbon content and decrease of hydrogen and oxygen. *Fig. 30* shows the two specific types of carbonaceous fuels – charcoal and coke. Both these fuels are produced by a thermal process – carbonisation (i.e. thermal decomposition in the absence of air). While charcoal is made from renewable biomass (mainly woody biomass), coke is produced from non-renewable types of bituminous coal. With regard to the distribution of fuels, wood charcoal, therefore, belongs to the category of renewable fuels and coke to the category of non-renewable fossil fuels. From the diagram in *Fig. 30*, it is clear that the carbonisation of woody biomass to charcoal and bituminous coal to coke decreases the H/C and O/C ratios. This is due to the removal of water, hydrogen, CO, CO_2 and some part of hydrocarbons in the process of carbonisation.

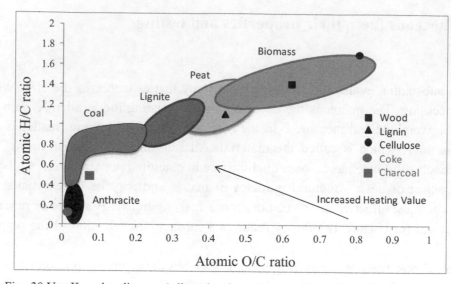

Fig. 30 Van Krevelen diagram (adjusted and supplemented by authors according to [11])

Fig. 31 shows the elemental composition of combustibles (energy bound in the fuel) of different types of carbonaceous fuels, which leads to the known fact (explained above) that the older fuel is geologically (with higher carbonisation), the more carbon and the less oxygen it contains.

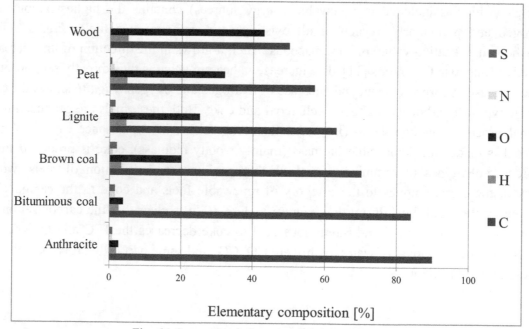

Fig. 31 Composition of combustible in various fuels

All the solid carbonaceous fuels contain the following essential components: combustible volatile matter (VM), water (W) and crude ash (A). Generally, any fuel composition can be expressed by equation (2):

$$VM + W + A = 1 \qquad (2)$$

A combustible forms that part of the fuel, which releases heat, i.e. chemically bound energy in the fuel, by oxidation. One part of the combustible consists of elements of carbon, hydrogen and sulfur. These are so-called active substances of the combustible, by oxidation of which heat is released in the so-called exothermic reactions. The second part of the combustible is made up of so-called passive substances, which do not deliver the heat (in the so-called endothermic reactions) but are bound to the organic matter of the fuel. Oxygen and nitrogen are the passive substances of the combustible. The combustible in carbonaceous fuel is also divided into non-volatile and volatile combustible. Non-volatile flammable substance is formed fixed carbon. The volatile combustible matter contains mainly gaseous hydrocarbons C_xH_y (g) (methane CH_4 (g) has the majority representation), CO, CO_2, H_2, H_2O (g), and H_2S. The composition of carbonaceous fuel is illustratively shown in *Fig. 32*.

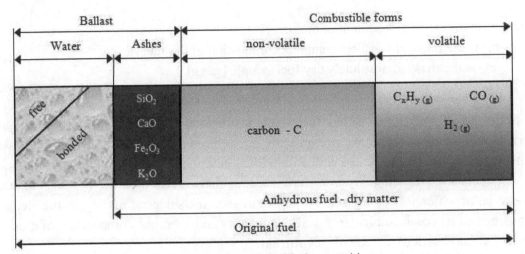

Fig. 32 Representation of fuel composition

Water and crude ash are called ballast. The presence of water, in any form, reduces the utility value of the fuel, as this component is a non-combustible portion of the fuel. Its content and form depend on the type of fuel, the mode of extraction, and the treatment and processing techniques [7].

Water content (or relative moisture) of the fuel can be calculated according to the equation (3).

$$W = \frac{m_1 - m_2}{m_1} . 100 \qquad (3)$$

where: W   is the relative moisture of the fuel [%],
        m_1 is the mass of a sample of raw fuel [g, kg],
        m_2 is the mass of the fuel samples after drying [g, kg].

Crude ash also belongs to the non-combustible component of the fuel and reduces its utility value. It is composed of minerals (i.e. oxides mix of SiO_2, CaO, Fe_2O_3, Al_2O_3, Na_2O and

K_2O), which are found in fuel. In relation to the crude ash, it is necessary to explain one more term – ash. These two terms are often used interchangeably or as identical in the technical as well as teaching and research practice. The crude ash is thus to be seen as an original part of the fuel (original material), which can be used as an energy or metallurgical (or chemical) raw material. The ash is the solid residue, resulting from the complete combustion of the fuel.

The ash content of the fuel can be calculated according to the equation (4).

$$A = \frac{m_p}{m_d} \cdot 100 \tag{4}$$

where: A is the crude ash content of the fuel [%],
$\quad$ m_p is the mass of ash after complete combustion of the fuel [g, kg],
$\quad$ m_d is the mass of absolutely dry fuel sample [g, kg].

The basic analysis of carbonaceous fuels includes determination of water, ash and volatile matter. The fixed carbon content is standardly not determined but is subsequently calculated (balanced) based on the mass loss in the individual determination (water and volatile matter) and residual ash. Extended analysis of carbonaceous fuels is based on the chemical nature of the material of which it is composed. Properties of such fuels depend mainly on the elemental composition of its combustible. However, the quantity and chemical composition of crude ash (or ash after combustion) are equally significant.

6.1.   Chemical and phase composition

In the previous part, it was said that each carbon fuel contains four basic elements: carbon, hydrogen, oxygen and nitrogen. These are the key elements of organic matter. The most important component of the organic material of carbonaceous fuel is carbon. Sulfur within a fuel is also a part of its combustible. Although sulfur is not considered an essential element in carbonaceous fuel, analysis of sulfur belongs to the basic quantitative assessment of the fuel because its presence is undesirable in metallurgical processes. The quantitative representation of given elements in the volatile combustible matter is different and is related to the geological age and degree of coalification [4, 5, 7]. Carbon may be present in the fuel usually only in the combustible matter (in non-volatile or volatile one, *Fig. 32*). Hydrogen is present in water and volatile combustible matter. Oxygen is also present in water and the volatile combustible matter, while it is a part of the crude ash as well. Nitrogen and sulfur are present in the volatile combustible matter. A part of sulfur can represent a constituent part of the crude ash. Organic matter, however, also contains traces of most of the known chemical elements in the form of different compounds.

A reference chemical analysis of some carbonaceous fuels is provided in *Tab. 4* for better understanding and comparison. Chemical analysis is divided into proximate analysis and ultimate analysis. Proximate analysis is focused on the determination of free water, ash, volatile matter and fixed carbon. Fixed carbon content is usually estimated by the loss in mass through the determination of free water, volatile matter and residual ash after combustion of carbon. The sum of components in the proximate analysis should be 100 %. The ultimate analysis is focused on the determination of ash content and analyses of elements C, H, O, N and S. The analysis is performed on a dried sample of the carbonaceous fuel. The oxygen content (O) is generally estimated by the analysis of ash, C, H, N and S. The sum of components within the ultimate analysis should be 100 %.

Tab. 4 Chemical analysis of carbonaceous fuels

Type of fuel	Proximate analysis – approximate analysis [wt%]				Ultimate – element analysis [wt%]					
	H_2O (W)	Ash (A)	Combustible volatile matter (VM)	Fixed carbon (C_{FIX})	Ash (A)	C	H	O	N	S
Sunflower husks	9.3	3.2	75.5	12.0	3.2	46.8	6.1	43.1	0.7	0.10
Wood sawdust	7.1	1.5	83.4	8.0	1.5	50.6	5.9	41.7	0.2	0.10
Wood	3.1	2.4	77.3	17.2	2.4	55.2	5.7	36.4	0.2	0.10
Charcoal	2.2	1.8	6.4	89.6	1.8	92.4	1.4	3.9	0.4	0.05
Brown coal	12.4	6.8	54.8	26.0	6.8	61.9	3.9	25.1	1.2	1.10
Bituminous coal	6.3	7.3	30.3	56.1	7.3	80.2	5.2	5.5	1.4	0.40
Blast furnace coke	3.5	10.5	0.8	85.2	10.5	87.2	0.3	0.5	1.1	0.40
Coke powder	5.5	12.1	1.5	80.9	12.1	85.4	0.3	0.6	1.3	0.30

In the context of modern methods, the method of activity measurement by scintillation spectrometry (LSC) is applied to determine the amount of fixed carbon in a sample of carbonaceous fuel. However, this methodology is also expensive, and therefore, the calculation of fixed carbon content is sufficient in the proximate analysis. In addition, the oxygen content can be determined within the ultimate analysis – e.g. quantometrically, similarly as C, H, N, and S.

Let us describe yet another method that enables understanding the composition of carbonaceous fuels. *Fig. 33* shows a thermogravimetric analysis – a course of the thermogravimetric curve (TG) of wood – by which it is possible to analyse the basic components of this carbonaceous fuel in the proximate analysis [24]. In the thermogravimetric curve, the release of water (1) at 105 °C, the evaporation of volatile matter (2) at 900 °C, and oxidation of fixed carbon (3) at 900 °C are recorded in sequence. The release of water and volatile matter take place without access of air (e.g. in an inert atmosphere of nitrogen or argon), and the oxidation of carbon is carried out in an oxidising atmosphere (e.g. air). After these thermal processes, the rest of the sample of analysed carbonaceous fuel forms ash (4). According to the thermogravimetric analysis in *Fig. 33*, the final composition of wood is as follows: 3 % of H_2O, 76 % of combustible volatile matter, 15 % of fixed carbon and 6 % of ash.

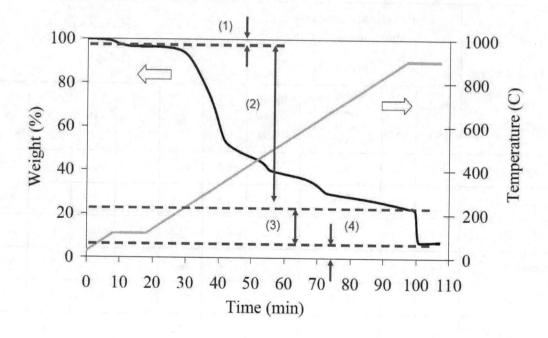

Fig. 33 Illustration of thermogravimetric analysis of carbonaceous material – wood [24]
(1) moisture (W),
(2) combustible volatile matter (VM),
(3) fixed carbon (C $_{FIX}$),
(4) ash

Tab. 5 shows the chemical composition of certain types of biomass and fossil fuels – within the ultimate analysis (without analysis of ash) [1].

Tab. 5 Chemical composition of biomass and fossil fuels [1]

Type of fuel	Chemical composition in a dry state [%]					
	C	H	O	N	S	Cl
Spruce wood with bark	49.8	6.3	43.2	0.13	0.015	0.005
Beechwood with bark	47.9	6.2	45.2	0.22	0.015	0.006
Poplar wood – short shoots	47.5	6.2	44.1	0.42	0.031	0.004
Willow wood – short shoots	47.1	6.1	44.3	0.54	0.045	0.004
Bark form coniferous wood	51.4	5.7	38.7	0.48	0.085	0.019
Rye straw	46.6	6.0	42.1	0.55	0.085	0.400
Wheat straw	45.6	5.8	42.4	0.48	0.082	0.190
Barley straw	47.5	5.8	41.4	0.46	0.089	0.400
Rape straw	47.1	5.9	40.0	0.84	0.270	0.470
Wheat grain with straw	45.2	6.4	42.9	1.41	0.120	0.090
Wheat grain	43.6	6.5	44.9	2.28	0.120	0.040
Rapeseed	60.5	7.2	23.8	3.94	0.100	-
Miscanthus sinensis	47.5	6.2	41.7	0.73	0.150	0.220
Agricultural hay	45.5	6.1	41.5	1.14	0.160	0.220
Pasture grass	46.1	5.6	38.1	1.34	0.140	1.390
Charcoal	90.1	1.2	2.4	0.50	0.005	0.020
Brown coal	65.9	4.9	23.0	0.70	0.390	0.100
Bituminous coal	72.5	5.6	11.1	1.30	0.940	0.100
Blast furnace coke	88.1	0.5	0.6	1.20	0.420	0.200

Water

Water is present in the fuel in various forms. It may be free but also bound in different forms. Free water is also called hygroscopic. It is removed by drying at 105 °C. Bound water may be present in the form of hydrates in the crude ash or can be bound to the combustible – i.e. constitutional water. The hydrate and constitutional water are removed at higher temperatures than hygroscopic water, typically in a wide temperature range of 150–700 °C.

Crude ash

Crude ash is formed as a result of reactions of minerals that are present in the carbonaceous fuel (e.g. coal or biomass) with oxygen. These minerals have been a part of fuel from its origins (primary minerals) or have been brought into the fuel (secondary minerals). Given the fact that crude ash contains various oxides (e.g. SiO_2, Al_2O_3, CaO, MgO, Fe_2O_3, K_2O and Na_2O), it is necessary to know not only their total amount and composition but also their phase and mineralogical composition. *Tab. 6* shows the chemical composition of crude ash in various carbonaceous fuels [25]. In the case of fuels from biomass (dendromass), the main represented component is calcium oxide (CaO), whereas in the case of fossil fuels (e.g. bituminous coal and coke), it is silicon dioxide (SiO_2) and alumina (Al_2O_3).

Tab. 6 Chemical composition of crude ash in carbonaceous fuels [25]

Type of fuel	Ash content [%]	Chemical composition of ash [%]						
		K_2O	Na_2O	MgO	CaO	P_2O_5	SiO_2	Al_2O_3
Beech	0.57	0.09	0.02	0.06	0.31	0.03	0.01	0.03
Birch	0.28	0.03	0.02	0.02	0.15	0.02	0.01	0.01
Oak	0.52	0.05	0.02	0.02	0.37	0.03	0.01	0.01
Pine	0.27	0.04	0.01	0.03	0.14	0.02	0.01	0.01
Spruce	0.29	0.03	0.02	0.02	0.15	0.02	0.01	0.02
Larch	0.29	0.04	0.02	0.07	0.10	0.03	0.01	0.01
Charcoal	4.70	0.20	0.05	0.50	2.50	0.01	1.20	0.12
Bituminous coal	7.50	0.10	0.03	0.20	0.54	0.02	3.82	1.85
Coke	10.62	0.12	0.04	0.20	0.72	0.01	5.23	2.31

Mullite ($Al_6Si_2O_{13}$), quartz (SiO_2), hematite (Fe_2O_3), calcite ($CaCO_3$) and hedenbergite ($CaFeSi_2O_6$) are typical compounds that are found in the ash of carbonaceous fuels. *Tab. 7* shows the phase composition of ash content in some carbonaceous fuels [26]. The phase composition of the ash is in accordance with the chemical composition of the ash of fossil fuel and dendromass, when the majority compounds in the fossil fuel (coke) ash are based on minerals mullite and quartz, and when the majority compounds in ash from dendromass (charcoal and oak sawdust) are based on minerals calcite, magnesium and calcium oxide. The high amorphous share in oak sawdust, which can be related to minerals in this type of wood biomass that are difficult to identify, is of particular interest.

Tab. 7 Phase composition of carbonaceous fuel ash [26]

Identified phase composition		Coke	Charcoal	Oak sawdust
Chemical formula	Mineral name	[wt%]	[wt%]	[wt%]
$(Ca_{0.94}Mg_{0.06})CO_3$	Calcite	-	57.3	-
$MgCO_3$	Magnesite	-	26.8	12.1
$Ca_6Mn_6O_{16}$	-	-	15.9	-
$Al_{1.25}Si_{0.75}O_{4.87}$	Mullite	50.4	-	-
$CaFeSi_2O_6$	Hedenbergite	4.3	-	-
$CaSO_4$	Anhydrite	8.0	-	-
SiO_2	Quartz	16.9	-	10.2
$Ca_2Fe_{1.54}Al_{0.46}O_5$	Brownmillerite		-	-
Fe_2O_3	Hematite	16.5	-	-
Fe_3O_4	Magnetite	4.9	-	-
MgO	Periclase	-	-	-
$CaCO_3$	Calcite	-	-	56.0
CaO	Calcium oxide	-	-	19.8
Amorphous share	-	**19.6**	**41.0**	**84.0**

Since carbonaceous fuels are used in thermal processes (combustion and reduction), it is necessary to know the melting temperature of the solid non-combustible residue, i.e. ash. The melting points of phytomass ash are about 850–1000 °C. For dendromass and fossil fuels, these temperatures are higher (1100–1450 °C). *Fig. 34* shows the high-temperature observation of ash from coke, charcoal and barley straw. The same temperature of 1000 °C was used for the comparison. At this temperature, various course of melting of the ash from these carbonaceous fuels was observed. While the ash from coke was almost unmelted, the ash from charcoal was only partially melted, and in the case of barley straw ash, a significant melting of the large part of the sample was observed. *Fig. 35* shows the high-temperature observation of a different sample of ash from charcoal with a relatively large melting interval (approximately 1040–1320 °C) for this carbonaceous fuel.

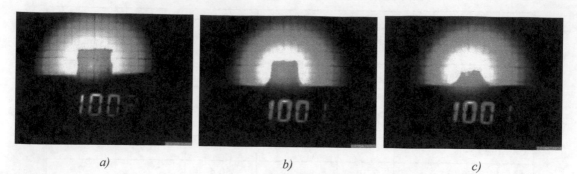

a)   b)   c)

Fig. 34 High-temperature observation of ash from carbonaceous fuels – at 1000 °C
a) coke ash, b) charcoal ash, c) barley straw ash

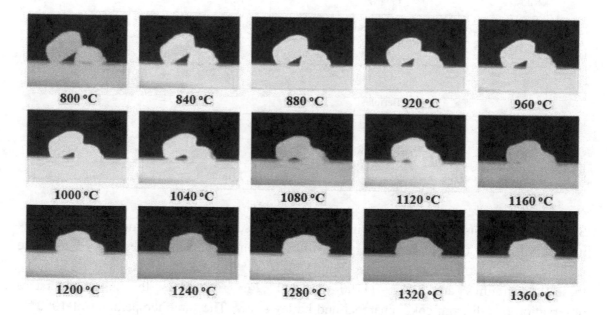

Fig. 35 High-temperature observation of charcoal ash

In *Tab. 8*, several types of biomass and fossil fuels are listed [1].

Tab. 8 Ash contents and melting points of various carbonaceous fuels [1]

Solid fuels	Ash content (A) [%]	Melting point of ash [°C]
Pasture grass	8.8	1150
Wheat grain	2.7	687
Rape straw	6.2	1273
Wheat straw	5.7	998
Rye straw	4.8	1002
Barley straw	4.8	980
Agricultural hay	5.7	1061
Beechwood	0.5	1350
Willow wood	2.0	1283
Poplar wood	1.8	1335
Spruce wood	0.6	1426
Bark from coniferous wood	3.8	1440
Brown coal	5.1	1050
Bituminous coal	8.2	1250
Anthracite	5.3	1340
Coke	11.0	1380
Charcoal	5.3	1280

Combustible

Combustible volatile matter (VM) of carbonaceous fuel is composed of carbon, oxygen, hydrogen, nitrogen and sulfur. Volatile matter (VM) is the part of combustible that escapes when fuel sample is heated without access of air to the temperature of 850 °C. The rest is a non-volatile share of the combustible – i.e. elemental fixed carbon (C_{fix}). In the early stages of volatile combustible release, water vapour and oxygen-containing gases (CO, CO_2) escape. In parallel to these processes, some oxygen-containing compounds are converted into volatile aldehydes, phenols and hydrocarbons, while CO, CO_2, methane, ethane and other gaseous products (e.g. H_2S and SO_2) are expelled further. These gaseous products evaporate at the temperatures of approximately 200–900 °C, while the major portion is released at the temperatures below 500 °C, when hydrogen evaporates intensively as well. Of course, the issue of the volatile combustible release from carbonaceous fuel is much more complex and is related to the chemical processes of thermal destruction of fuels [27]. In terms of the chemical composition of fuel, it is very important to understand the processes of the release of individual gaseous components during combustion processes. These gaseous constituents of volatile combustible are involved in chemical reactions and influence the resulting thermal effects of fuel combustion.

Sulfur

Sulfur is present in organic and inorganic materials of carbonaceous fuels as pyritic, sulfate, elemental and organic sulfur. Although sulfur improves combustion heat and calorific value of the fuel, it is always the unwelcome component of the fuel. All sulfur compounds undergo certain changes during thermal processing. During combustion, sulfur is burned to SO_2, which evaporates with the flue gas into the atmosphere, where it is extremely damaging to vegetation and other living organisms. At lower temperatures, it combines with water or steam and intensely corrodes lining of chimneys and steel parts of equipment. A part of sulfur from carbonaceous fuel is concentrated in the metallurgical products (iron and steel), which considerably reduces the value of high-sulfur fuel for metallurgical purposes.

Phosphorus

Phosphorus is always present in any carbonaceous fuel at the level of hundredths to tenths of a percent. It is not much indeed, but even such a small amount is very problematic. Phosphorus originates from coal-forming plants, and its content ranges from 0.01 to 0.3 %. Phosphoric salts do not decompose by burning or carbonization. Therefore, entire phosphorus content passes into the ash or metallurgical products. Phosphorus is undesirable in the carbonaceous fuels used for the production of special types of iron, quality ferroalloys, calcium carbide, etc.

6.2. Petrographic composition

The structure of carbonaceous fuels is very complex and cannot be described merely by chemical elements it comprises. It depends on the level of coalification (in the case of fossil fuels) and the nature of formation and growth (for biomass). The macroscopic and microscopic state of this material can be explained on the example of the study of coal. The science or discipline that deals with describing composition and properties of rocks is called petrography. Petrographic properties of coal are defined on the basis of macroscopic and microscopic study [7, 28, 29, 30].

Macroscopically observable components of coal are referred to as lithotypes. They are formed by parallelly ordered carbonaceous matter (i.e. stripy texture structure). A maceral is an important term in coal petrography. The maceral is a microscopically distinguishable component of coal. It is also referred to as microlithotype. The macerals are smallest particles that are distinguishable under a microscope, and thus can be considered an analogy of minerals in the rock. In the microscopic picture, they can be distinguished based on such characteristics as colour, gloss or light reflection (reflectance), morphology (*Fig. 36*), grain size and the strength. The macerals are also specified as the individual components of an organic material having defined chemical and physical properties.

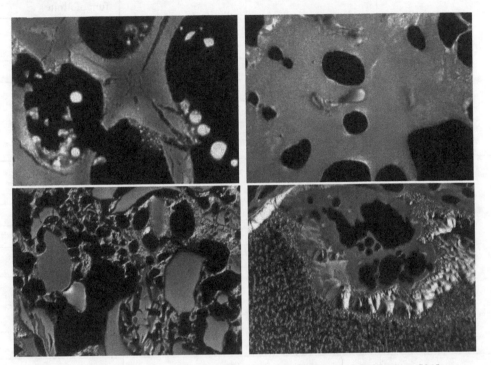

Fig. 36 Morphology of the highest quality coal in the world – Blue Gem [31]

We know three groups of macerals: vitrinite, liptinite and inertinite. In the coal of the same level of coalification, vitrinite usually has relatively higher oxygen content, liptinite contains more hydrogen, and inertinite more carbon. Liptinite contains a relatively large proportion of volatile components – hydrocarbons. Some macerals represent the remains of plants from which they originated and whose structure has been preserved in coal. Other macerals are products of degradation of organic parts that have changed significantly, and their origin is nearly impossible to determine. If there are differences in the origin of macerals (e.g. different form or structure), then we distinguish maceral types. If the structure of macerals is assigned to a particular part of a plant, the term maceral variety is used. The system of bituminous coal macerals classification is given in *Tab. 9*.

The coal petrography is relatively complex but has a very important industrial use. In metallurgy, it is also used to evaluate the characteristics of metallurgical coke and its application utilisation. It is used to review the degree of oxidation of coal and control the combustion properties of coal. When evaluating coal for coke production, microscopic research can be used to assess the presence of individual maceral components or content of inorganic components. Simply put, just using petrographic study (besides other analyses) it is possible to exclude certain types of coal from the energy and industrial processing.

Tab. 9 Stopes-Heerlen's system of bituminous coal maceral classification [28]

Maceral group	Maceral	Submaceral	Maceral variety
Vitrinite	tellinite	tellinite 1	cordaitotelinite
		tellinite 2	fungotelinite
			xylotelinite
		telocollinite	lepidophytotelinite
		gelocollinite	sigillariotelinite
	collinite	desmocollinite	
		corpocollinite	
	vitrodetrinite		
Liptinite	sporinite		tenuisporinite
			crassisporinite
			microsporinite
			macrosporinite
	cutinite		
	resinite		
	alginite		pila-alginite
	liptodetrinite		reinschia-alginite
Inertinite	micrinite		
	macrinite		
	semifusinite		
	fusinite	pyrofusinite	
		degradofusinite	
	sclerotinite	fungosclerotinite	plectenchymite
			corposclerotinite
			pseudocorposclerotinite
	inertodetrinite		

Vitrinite

Vitrinite is the most frequently and most abundantly represented group of macerals in coal. Vitrinite macerals are identified in coal with higher coalification. Vitrinite almost linearly changes its physical and chemical properties depending on the degree of coalification. It is a chemically reactive component of coal. It releases volatiles and produces sufficient amount of the liquid phase. Vitrinite may be easily distinguished in the microscope.

Liptinite

Liptinite macerals are the less represented component of coal. Depending on coalification, liptinites are chemically reactive and contain a higher proportion of hydrogen and volatiles. During coal combustion, they increase the content of tar components. They are an important indicator for assessing the possibility of using the coal for gasification to produce gases with a high content of hydrocarbon components – mainly methane.

Inertinite

Inertinite macerals are present in coal in varying amounts, usually less than the vitrinite. Macerals of this group have high reflectance, a relatively high carbon content, low hydrogen and volatiles, and a higher content of aromatics. They are considered chemically inert [32]. The heterogeneity of the chemical composition and structure of coal causes different light reflectance of maceral groups. The differences are so significant that the reflectance of maceral groups is their basic diagnostic feature. The changes in reflectance are also caused by varying degrees of coalification. In Fig. *37–38*, there are the textures of the Blue Gem and Staszic coal. It is obvious that in the coal matter, maceral components are commonly ingrown with mineral – often clay particles. The most common objective of the microscopic study of coal is type determination from the quantitative representation of microlithotypes (macerals) to be used for describing coal and determination of its quality – also in view of its possible use.

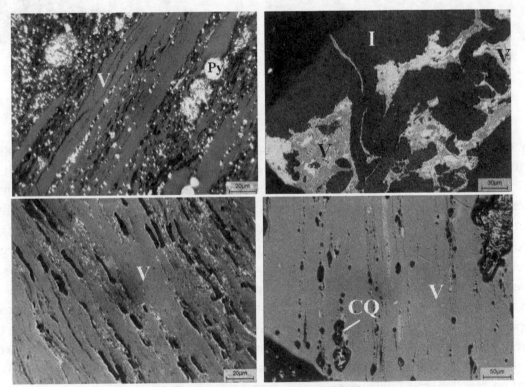

Fig. 37 Texture of Blue Gem coal [31]
(V – vitrinite, Py – pyrite, I – clay particles, CQ – quartz)

The analysis and evaluation of maceral composition of carbonaceous matter is a very complex scientific discipline. The authors of this publication wanted to point out these specific (maybe not very well known and used) characteristics that distinguish different fuels, as well.

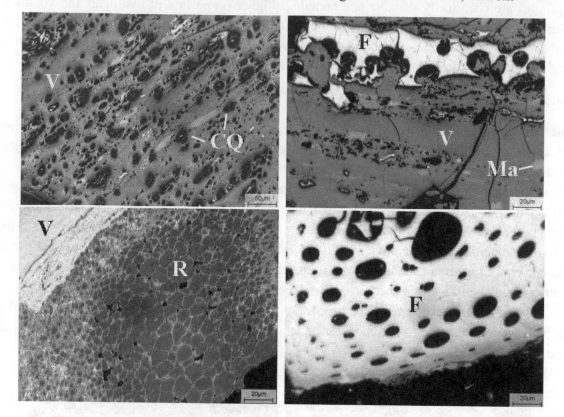

Fig. 38 Texture of Staszic coal [31]
(V – vitrinite. F – fusinite. Ma – macrinite. R – resinite. CQ – quartz)

Why is it important to study and analyse maceral composition? Why is it necessary to know the structural composition of coal? Is the basic chemical analysis of carbonaceous fuels not sufficient? We can find the answer to these questions only if we understand the relationship between the structure (microtexture) and physical-chemical properties of carbonaceous fuels. Reactivity is one of the most important properties of coal and coke used in metallurgical processes. Different reactivity of carbonaceous fuels is required for various metallurgical processes. For instance, blast furnace coke should be less reactive, while the coke for ferroalloy production should be as reactive as possible. In general, the reactivity of coal and coke decreases with the increasing degree of coalification of coal. In terms of maceral composition of coal, vitrinite is the active maceral in coal, while inertinite is the inactive one. Microtexture of coal and coke affects primarily their reactivity. Coarse anisotropic components (optically active component with a greater surface area) are less sensitive to the oxidation reaction and temperature shocks than fine anisotropic and isotropic components. This knowledge is also used in the evaluation of carbonaceous fuels in their metallurgical application. ***Tab. 10*** shows a generalised influence of the structure on the reactivity of coal, which has been discussed above [31].

Tab. 10 Impact of the structural composition of coal on its reactivity (reference)
[modified by authors according to 31]

	Light reflectance ($R_{o\,max}$) [%]	Texture	Texture – photo	CO_2 reactivity
V I T R I N I T E	< 0.8	isotropic		1 (higher)
	0.9–1.1	fine mosaic		3
	1.2–1.4	coarse mosaic		4 (lower)
	1.5–1.8	coarse anisotropic		4 (lower)
	> 1.9	isotropic		2
INERTINITE		inert		2–3

Legend to ***Tab. 10***: Evaluation of reactivity was carried out based on the order in the reactivity test.

6.3. Physical and mechanical properties

Colour, appearance, shape, weight, particle size, porosity, temperature, specific weight: these are the physical properties that characterise every solid matter. The relationship between the size of the bodies, grains and particles of solids on one hand, and their utility properties on the other hand, is clear. Many physical-chemical properties of solids are considerably influenced exactly by particle size. For many materials, the size distribution is, therefore, one of the most important properties and compliance with a certain optimal composition significantly affects their processing. This is one of the reasons why there are generally very strict requirements for the most accurate characterisation of grain composition of solids. Other physical properties of substances, e.g. specific weight and the surface area, also depend on grain composition. Let us consider two extreme cases of the solid substance composition. Ideal (homogeneous) state – the substance may contain particles of the same shape and approximately same size, *Fig. 39a*. Non-homogeneous state – the substance contains particles, whose shape and size are in the range of several measurement units, *Fig. 39b*. Now suppose that, based on a model example in *Fig. 39*, a carbonaceous fuel (coal, coke or biomass) may have different grain size distribution. If we use such fuel, which has a non-homogeneous state, e.g., for combustion, we will achieve significantly worse thermo-technical parameters of the combustion process. Also, reducing conditions, using a carbonaceous reducing agent, will be considerably worse in the case of its non-homogeneous state. This is due to the thermodynamic and particularly kinetic conditions of processes where the carbonaceous substance is used. These will be explained in the following chapters on the application utilisation of carbonaceous fuels.

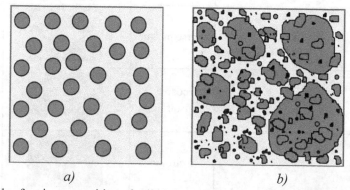

a)   b)

Fig. 39 Example of grain composition of solid substance – a) ideal state, b) non-homogeneous state

Some physical properties of carbonaceous fuels can be only assessed by visual observation. Such properties are, for example, coal colour and appearance. The colour of coal is influenced by the origin of coal, the degree of coalification, petrographic composition and content of minerals. Coal is divided in brown and bituminous coal by colour. The appearance of coal is assessed based on brightness, which is given by the ability to reflect light. Brightness depends on the petrographic composition, degree of coalification and crude ash content. The fracture, cleavability and shape of coal are other physical properties. The fracture of coal can be conchoidal, prismatic, cleavage, earthy, woody and crumbly.

Grain size composition and specific weight of carbonaceous fuels

Granularity (grain size composition) is evaluated according to the results of sieve analysis, from which the mean grain size of carbonaceous fuel is calculated as well. Currently, there are many evaluations and interpretations of the results of grain size composition of granular materials. When the grain size composition of solid carbonaceous fuels is determined, percentages of various grain size classes and their cumulative shares are standardly calculated. *Fig. 40* shows the grain size composition of the two carbonaceous fuels (coke dust and sawdust). It is clear just from the macro photograph that for both materials the particle size is at the level of tenths of a millimetre up to one millimetre. If we look closely at the percentages of the various grain size classes and their cumulative shares, we ascertain that in the case of coke dust, approximately 21 % is in grain size class of 0.3 to 0.5 mm and approximately 65 % of this fuel is below 0.5 mm. In the case of sawdust, almost 80 % of the fuel is below 0.5 mm. It follows from this demonstration that each fuel has different and distinct grain size composition.

a)

b)

Fig. 40 Grain size composition of a) coke powder, b) sawdust from oak

Density is one of the important physical properties of carbonaceous fuels. It has been already mentioned that the density is also affected by grain size, but it is strongly influenced by the moisture as well. The specific weight of fossil fuels varies due to the grain size composition, quantity and composition of crude ash, original moisture content, the degree of coalification or as a result of petrographic composition. The specific weight of biomass depends on the original moisture content, the amount of volatile combustible and the size of individual grains. The amount and composition of the crude ash have mostly a minimum impact on the density of biomass – given its low content. Conversely, the moisture of biomass is a parameter that significantly affects its specific weight. Up to three density values are typically analysed for carbonaceous fuels – apparent, bulk and real density.

The real density of fossil fuels is about 1100–1800 kg.m^{-3}. These values are lower for various types of biomass, and as already said, they are highly dependent on moisture. The real density of biomass is about 400–900 kg.m^{-3}.

Porosity

Carbonaceous fuels have a porous texture. The porous texture, in general, can be characterised by an inner surface and pore size. The inner surface and its accessibility for the reactants have a determining effect on the speed of such processes as gasification, combustion, reduction and dissolution. The pores in the carbonaceous material are divided into the following groups: micropores, mesopores and macropores. Most of the internal surface is formed by fine pores, *Fig. 41*.

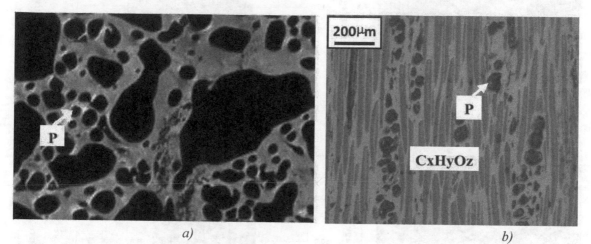

a)   *b)*

Fig. 41 Microstructure of the highest quality coal in the world – Blue Gem (a) and sawdust (b) [31. 33]
P – pores. C$_x$H$_y$O$_z$ – hydrocarbon group with bound oxygen

The porosity of the solid fuels is given by the ratio of pore volume to the apparent volume of the substance, and is calculated from the real density and the apparent density by the formula (5):

$$P = \frac{\sigma_s - \sigma_z}{\sigma_s}.100 \qquad\qquad (5)$$

where: P   is the porosity [%],
       σ_s is the real density [kg.m^{-3}],
       σ_z is the apparent density [kg.m^{-3}].

In addition to the real porosity of solid fuels, so-called open porosity is also determined, where pores extend up to the surface of the substance and are interconnected.

Mechanical properties of carbonaceous fuels

The most important mechanical properties of solid fuels include strength, brittleness, grindability and hardness. These properties are mainly related to fossil fuels and are essential for the treatment of these fuels by crushing and grinding. Strength is characterised by resistance to deformation and disintegration under the influence of external forces. Among solid fossil fuels, anthracites have the highest strength and coal of medium coalification degree has the lowest strength.

The brittleness is the ability to crumble under mechanical action with no plastic deformation. The brittleness of, e.g., carbonaceous fossil fuels depends on the degree of metamorphism, the petrographic composition, the content of the internal cracks and the homogeneity.

The grindability is expressed by the amount of work expended on creating a new grain surface by crushing in various types of mills or by the ability to create a certain granulometric composition after passing a milling unit. The grindability applies mainly to coal and is determined as a Hardgrove Grindability Index (HGI). If the HGI < 60, coal is difficult to grind. If the HGI > 60, coal is well grindable [7]. This property is crucial for coal that is used, e.g. for injection into a blast furnace. The comparative coal is bituminous coal from the Somerset mine in the US, with grindability equal to 100. The grindability is significantly influenced by the degree of coalification, the content of volatiles and the maceral composition of carbonaceous fuels.

The hardness of carbonaceous solid fuels is assessed by microhardness testing and represents the resistance that solid fuel exerts towards external mechanical action. The microhardness is evaluated according to the depth of indentation by the defined body into polished coal samples. Lignite has microhardness from 60 to 170 MPa, bituminous coal from 142 to 348 MPa, and gas coal have the lowest microhardness from 140 to 180 MPa.

Unlike fossil fuels (bituminous coal, anthracite and coke), in the case of biomass, standard mechanical properties of the original samples are not determined (except for charcoal, *Fig. 42a*). As these are mostly fine-grained materials of dendromass and phytomass, for example, strength, hardness and grindability are not determined on these samples. Nevertheless, the strength may be determined on solidified samples, which are prepared by briquetting (at a pressure of approximately 30 MPa). Briquetting provides not only a higher density (approximately 1200 $kg.m^{-3}$), but also higher calorific value. Moreover, these briquettes and pellets are characterised by unlimited storability and dustlessness, *Fig. 42b–c*.

Fig. 42 Types of biomass – a) charcoal, b) wood briquettes, c) wood pellets

6.4. Thermotechnical properties

Higher calorific value, lower calorific value, specific heat, thermal conductivity, thermal stability, ignition temperature and thermal expansion coefficients are included among thermo-technical properties of solid carbonaceous fuels. In terms of energy use, the most important characteristics of fuel are, inter alia, combustion heat, calorific value and ignition temperature.

Higher calorific value HCV [MJ.kg^{-1}] is the heat released by the complete combustion of 1 kg of fuel to CO_2, SO_2 and liquid water. Lower calorific value LCV [MJ.kg^{-1}] is the heat released under the same conditions, with the difference that instead of liquid water, water vapour is released. Lower calorific value (LCV, CV) is calculated by subtracting the heat of vaporisation of water from the higher calorific value. Water released by combustion is the sum of the water contained in the fuel (fuel moisture) and water formed by burning of hydrogen in the volatile combustible matter. The water content significantly affects the calorific value of fuel not only by reducing the solids content but also by energy consumption for its evaporation. Therefore, it is important to allow for the moisture of solid fuels in their purchase. Lower calorific value can be determined by formula (6):

$$LCV\,(CV) = HCV - 2.453\,.\,(w + 9.H_2) \tag{6}$$

where: LCV is the lower calorific value of fuel [MJ.kg^{-1}],

HCV is the higher calorific value of fuel [MJ.kg^{-1}],

w is the moisture content of fuel [kg.kg^{-1}],

H_2 is the hydrogen content of fuel [kg.kg^{-1}].

Tab. 11 shows calorific values of some types of fossil fuels and some types of biomass (fuels are classified according to the increasing heating value) [1]. These calorific values are given at zero moisture. It is clear from *Tab. 11* that the calorific value of dry wood is

18–20 MJ.kg^{-1}. Dry cereal straw and sawdust have the calorific value in the range of 16.5–17.5 MJ.kg^{-1}. In fact, the biomass always contains at least 10 % of water. When burning, the water evaporates, and it reduces the calorific value of the biomass solids. The calorific value of carbonaceous fuels is also reduced with time – due to the action of microorganisms, fungi and moulds.

Tab. 11 Calorific value and crude ash content of various carbonaceous fuel (dry state) [1]

Solid fuels	Calorific value CV [MJ.kg^{-1}]	Ash content A [%]
Pasture grass	16.5	8.8
Wheat grain	17.0	2.7
Rape straw	17.1	6.2
Wheat straw	17.2	5.7
Rye straw	17.4	4.8
Barley straw	17.5	4.8
Agricultural hay	17.6	5.7
Beechwood	18.4	0.5
Willow wood	18.4	2.0
Poplar wood	18.5	1.8
Spruce wood	18.8	0.6
Bark from coniferous wood	19.2	3.8
Brown coal	20.6	5.1
Bituminous coal	25.4	8.2
Anthracite	27.8	5.3
Coke	30.5	11.0
Charcoal	31.6	5.3

Specific heat is the amount of heat acquired by the unit volume of the material when heated by 1 K. Specific heat of peat is 1965 [J.kg^{-1}.K^{-1}], lignite from 1392 to 1442 [J.kg^{-1}.K^{-1}], bituminous coal from 859 to 1474 [J.kg^{-1}.K^{-1}]. The value of the specific heat increases with the content of volatile substances.

In *Tab. 11*, calorific values of some types of fossil fuels and biomass at zero moisture (in the dry state) are given. *Tab. 12* provides calorific values of wood and wood bark, depending on the water content [1]. It is clear that the moisture of biomass will have a significant impact on its thermo-technical properties.

Tab. 12 Real calorific value of wood and bark, depending on the water content [1]

Water content [%]	Calorific value CV [MJ.kg^{-1}]	
	Wood	Bark
0	18.5	18.8
10	16.4	16.7
20	14.3	14.6
30	12.2	12.5
40	10.1	10.5
50	8.0	8.4
60	6.0	6.3

The ignition temperature of solid fuels is essential for its association with the flammability and spontaneous combustion of fuels. The ignition temperature depends on the origin of fuel, its grain size, reactivity, moisture and degree of metamorphism. It changes with the volatile matter content with all fossil carbonaceous fuels. As the volatile matter content increases, the ignition temperature decreases, *Tab. 13*. Biomass fuels have low ignition temperatures.

Tab. 13 Ignition temperatures of different types of solid carbonaceous fuels
[modified by authors according to 1]

Solid fuels	Temperature of ignition [°C]
Wood – soft	220
Wood – hard	230
Dry peat	225–280
Brown coal	200–240
Bituminous coal	220–250
Anthracite	340–600
Charcoal	485
Semi-coke	340–380
Petroleum coke	410–430
Pitch coke	544–582
Bituminous coke	520–630

6.5. Other specific properties

In the previous chapters, we have become more familiar with the basic properties of solid carbonaceous fuels. Carbonaceous fuels, however, may perform functions other than their energy use. In the third chapter, it was stated that in the production of ferroalloys, coal and coke are used as majority carbonaceous reducing agents. Coal, in addition to the direct metallurgical use, is used to produce valuable metallurgical coke or semi-coke. The produced metallurgical coke is an important reducing agent in the blast furnace. Moreover, all of the carbonaceous substances can also be used as carburising agents (to increase the carbon content of certain metallurgical products – e.g. the carbon steel). Due to these "non-energy" purposes, it is necessary to know other specific properties of carbonaceous substances in the normal operating practice.

Electrical properties of carbonaceous reducing agents are relevant for their use in electric arc furnaces (e.g. in the production of ferroalloys). The most important of these characteristics include electric conductivity or its inverted value – specific electric resistance. In terms of electrical properties, lignite and bituminous coal can be described as semiconductors. The values of electric conductivity are in the range from 10^{-2} to 10^{-10} $\Omega^{-1}.m^{-1}$. The values of electric conductivity depend on the temperature, moisture, coalification degree, crude ash content, petrographic composition. For example, anthracite particles have several times lower specific electrical resistance than bituminous coal.

In the context of the use of coal for coke or semi-coke production, their coking properties are essential. Coking properties are manifested as changes in the carbonaceous matter at temperatures of 350–900 °C, which first softens when heated without air, then turns into a solid, sintered coke residue. Coking properties include caking, plasticity, dilation and swelling. They also include expansion and contraction of coal (or coal charge) and the course of degassing. All these features should be analysed to determine the size of the plastic zone, dilatation and contraction of the charge and semi-coke shrinkage – to produce high-quality coke.

6.6. Methodologies and testing

Trials, testing, analysis and comparison – these are the operations, without which we would not know any of the important properties of solid carbonaceous materials. The description and knowledge of the most significant methodologies that are currently used to analyse carbon fuels constitute the basic know-how in the field of technologies of iron, steel and ferroalloys.

There is currently a large number of methods to determine the properties of carbonaceous fuels. In *Tab. 14*, the most frequently used methodologies for testing of carbonaceous fuel are given. This overview makes it clear that the testing and analysis of carbonaceous fuels are very difficult and that it is necessary to perform a large number of tests to obtain a comprehensive view of the particular fuel. Inconsistent international standards that describe various methods of testing of solid carbonaceous fuels with certain modifications pose an issue as well. In testing of these fuels, it is also essential to correctly interpret the obtained results. Last but not least, it is needed to carry out tests and analyses regularly on a large number of samples, because carbonaceous fuels are very inhomogeneous. Analysis and testing of solid carbonaceous fuels are ones of the most difficult testing methods and the results of this activity are crucial for the assessment and decision on the further use of these fuels in the production of pig iron, steel and ferroalloys.

Tab. 14 Methods for testing of solid carbonaceous fuels

Methodology	State of tested sample	Characteristics or description of the methodology	Standard
Screening test for the determination of granulometric composition	original	sieve analysis 	STN 44 1340
Determination of the apparent bulk and real density of coal	original	calculations based on experimental tests	STN 441324
Determination of free water content	original	drying at 105 °C. atmosphere – air	STN ISO 589
Determination of ash	grainy	calcination at 830 °C. atmosphere – air	STN ISO 1171
continued on next page			

Determination of ash melting point (melting interval)	grainy	high-temperature observation at temperatures of 800–1450 $^{\circ}$C	STN 441359
Determination of volatile combustible matter (VM)	grainy	calcination at 850 $^{\circ}$C. atmosphere – inert	STN ISO 562
Determination of C, H, O, N, S	grainy	quantometric analysis	PP1-ZT21
Determination of sulfur content	grainy	analyser LECO	ASTM D4239
Maceral analysis	original, cut	petrographic analysis in a microscope	ISO 7404-3 GOST 9414.3-93
Determination of higher calorific value	grainy	calorimetric method in a pressure vessel	STN ISO 1928
Determination of lower calorific value	grainy	calculation after calorimetric method	STN ISO 1928
Determination of caking of coal	grainy	test by Damm, Kattwinkel, Rog	ASTM D 2639
Determination of plasticity of coal	grainy	test by Gieseler	ASTM D 2639-90
Determination of dilatation of coal	grainy	test by Arn	STN ISO 349
Determination of grindability	original	Hardgrove Grindability Index (HGI)	ASTM 409
Determination of mechanic strength of coke	original	drum test MICUM	STN 441327
Determination of reactivity index of coke (CRI)	original	determination of loss of mass after reaction with CO_2 at 1100 $^{\circ}$C	ISO/DIS 18 894
Determination of strength of coke after reactivity test (CSR)	after CRI test	determination by drum test	ISO/DIS 18 894

7. Use of traditional fuels in the production of agglomerates

The agglomerate is a basic input material for the production of pig iron in the blast furnace and plays an important role in the integrated metallurgical cycle. Pursuant to the forecasts of iron and steel production for the next decade, the continued need for Fe sinter production can be predicted. It sufficiently addresses the issue of substituting the lack of rich iron ore lump as a charge for the BF production. Furthermore, it addresses the issue of iron ore concentrate processing, and last but not least, the issue of processing different types of waste that are difficult to processes or unprocessable by another technology [34–37]. This fact predetermines the necessity to address and improve the technology of agglomerate production, analyse and explore the properties and the effect of different types of agglocharge on the overall sintering process, and the quality and metallurgical properties of sinter. It is also crucial to analyse the use of traditional carbonaceous fuels in this process and implement the intensification elements into the production of Fe agglomerate.

The agglomeration process is a summary of physical, physicochemical, chemical and thermal processes that alter the structure and composition of the input materials. This process occurs in a heterogeneous system of the gas-liquid-solid phase. The gas phase ensures the running of essential processes (fuel combustion, heat transfer and oxidation-reduction processes) [34, 35, 38]. The liquid phase is formed by melting of fine-grained particles and surface melting of larger particles. After the solidification, it provides the connection of individual particles in a porous sinter – agglomerate. In the agglomeration process, the entire charge does not melt. The liquid phase forms only in micro and macro volumes of individual grains. The agglomeration process is one of the ideal thermal processes in metallurgy. Low consumption of carbonaceous fuel (approximately 2.8 to 3.5 %) and the efficient recovery of heat provides heating of the material in a narrow strip to the maximum temperature of about 1380–1450 °C (standard temperature for this process is approximately 1280–1350 °C). The agglomeration process is largely an oxidation process, which is based on the combustion of carbon with oxygen forming CO and CO_2. Carbon is delivered into the agglomeration process through carbonaceous fuels – mostly in the form of fine powder coke, which is produced by the screening of metallurgical coke.

In the above text, the general characteristics of the agglomeration process was specified. By this technological process, ferriferous agglomerate (Fe agglomerate) is produced most often, which is used for the production of pig iron in the BF. Besides ferriferous agglomerate, other types of agglomerates are also produced. In the production of manganese ferroalloys in the EAF, manganese agglomerate is used as an important component of the charge. To protect BF hearth lining, ferriferous agglomerate with a higher content of titanium, also called titanium agglomerate, is produced. Within the agglomeration process technology, production of metallised agglomerate from secondary ferriferous materials (steelmaking dust and sludge) was also verified. In the production of all these agglomerates, fine-grained coke powder is mainly (mostly exclusively) used as a fuel.

7.1. Fe sinter production

Ferriferous agglomerate or sinter is one the main ferriferous charge components in the production of pig iron in the blast furnace. It is produced by high-temperature sintering of fine iron ore, iron ore concentrates and other ferriferous materials (e.g. secondary materials from iron and steel production). However, the process of iron ore sintering allows operating only with such amount of wastes (e.g. blast furnace and steelmaking sludge and dust, turnings, chips, etc.) that is technologically and economically acceptable.

Sinter plant generally consists of two basic sections – cold and hot section. Cold section of charge preparation includes tippers, equipment for crushing, grinding, ore storage, homogenization dumps, mixing tanks and pre-pelletisation drums. The cold section is followed by hot manufacturing section, which includes the agglomeration (sintering) strand with exhaust gas capturing equipment and agglomerate cooler.

The cold section ensures supply of raw materials, granularity adjustment to the required lumpiness, averaging the chemical composition of materials, and pre-pelletisation of final agglomeration charge. If we look at the material flow of input raw materials in more detail, we would find out that agglomixture from homogenization stocks (outdoor and indoor) contains a metal-bearing mixture (aggloore and concentrate), defined basic components, part of the fuel and ballast. This agglomixture is transported along transport routes where return agglomerate (below 5 mm) is added to it, *Fig. 43*.

Thus, adjusted mixture is mixed and wetted at the second stage of mixing and transported to the final processing. It is followed by mixing at the third stage, where the required amount of agglomeration fuel and water is added to the agglomixture that is turned into the finished agglomeration charge. In addition to the required chemical composition, the agglomeration charge should have the required particle size distribution with proper moisture and complying fuel content.

The task of the hot production section is to produce the agglomerate of the required quality from delivered raw materials (in the form of agglomeration charge). This section provides the ignition of the agglomeration charge, sintering of agglomeration charge, discharge of hot flue gas (and its following cleaning), cooling of the agglomerate, sorting, and transport of agglomerate to the blast furnace. The main criterion of the agglomeration process is the quality produced agglomerate while maintaining the ecological nature of the production. A comprehensive technological scheme for the preparation and processing of agglomeration charge, i.e. Fe agglomerate production, is shown in *Fig. 43*.

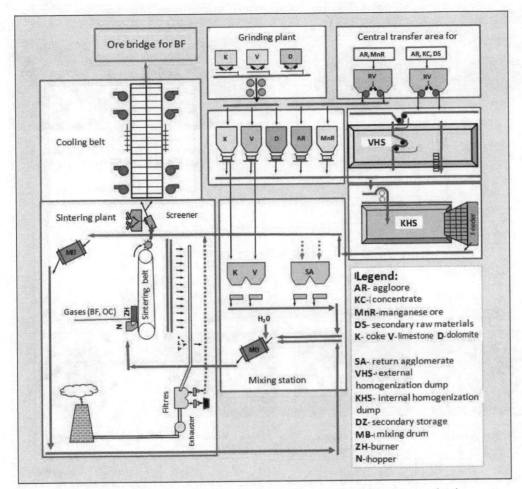

Fig. 43 Comprehensive technological scheme for agglomerate production

For a comprehensive evaluation of the agglomeration charge, it is necessary to examine the properties of the chemical composition, mineralogical composition and physical properties of individual components of the charge. Since the agglomeration charge is of a polycomponent nature, it is essential to analyse the impact of its individual components on the final properties of the resulting agglomerate.

Fig. 44–46 show the individual input raw materials for agglomeration process and their concise specifications.

Fig. 44 Input ferriferous raw materials

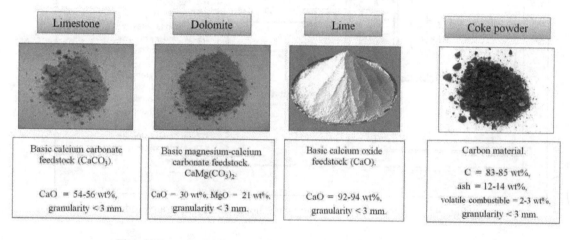

Fig. 45 Input basic components   Fig. 46 Fuel

Fig. 47 shows a flow diagram illustrating the key equipment of agglomeration operation – sintering belt (strand). Contemporary sintering plants usually operate on the basis of Dwight-Lloyd type strand. The apparatus consists of a set of grate type sintering boxes, which are connected and moving.

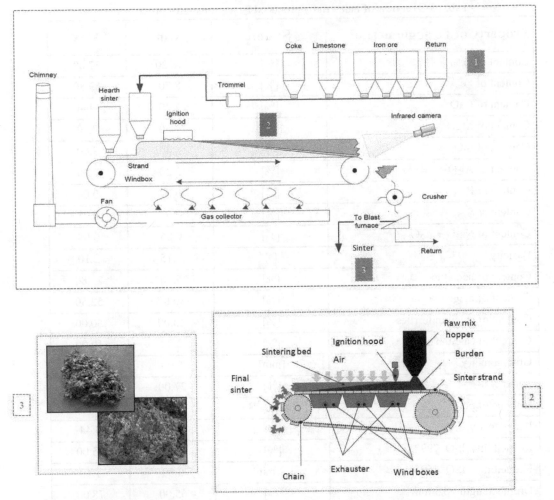

Fig. 47 Technological scheme of Fe sinter production plant of Dwight-Lloyd type [according to 39]
1) input materials, 2) scheme of Dwight-Lloyd sinter strand, 3) Fe agglomerate

Input charge is delivered to the sintering strand by a drum feeder. The integral parts of the sintering strand are wind boxes and exhausters for sucking air through the layer of agglomeration charge. Ignition of the top layer of charge is done by external heat source ignition hood burners (natural gas, blast furnace gas and coke oven gas are used the most). At the end of the strand, the sintered hot agglomerate is crushed and transported, after separation of undersize share (below 5 mm), to cooling strand, and then to blast furnace stocks (storage bins). The properties of produced agglomerate are determined by the requirements of blast furnace technology and consequently its production technology. Chemical, physical, mechanical and metallurgical properties constitute the essential requirements for the blast furnace agglomerate. In **Tab. 15**, some important properties of industrially produced agglomerates are listed, while the critical requirement for all agglomerate properties is stability.

Tab. 15 Overview of industrially produced Fe agglomerates properties

Property of Fe agglomerate	SI unit	Min	Max
Content of Fe_{TOT}	[%]	48.20	57.00
Content of FeO	[%]	8.70	19.80
Content of CaO	[%]	7.30	14.40
Content of MgO	[%]	4.50	7.10
Content of SiO_2	[%]	5.40	9.60
Content of Al_2O_3	[%]	2.20	2.50
Content of P	[%]	0.02	0.04
Content of S	[%]	0.03	0.05
Content of $Na_2O + K_2O$	[%]	0.05	0.08
Basicity	[-]	1.15	2.10
Content of hematite	[%]	5.40	27.80
Content of magnetite	[%]	30.00	52.50
Content of calcium ferrites	[%]	20.00	50.00
Content of calcium silicates	[%]	2.00	10.00
Granulometry	[mm]	5.00	50.00
Porosity	[%]	27.00	38.00
Volume of pores	$[mm^3.g^{-1}]$	23.05	41.25
Real density	$[g.cm^{-3}]$	4.13	4.44
Reducibility (ISO 7992)	[%]	60.00	85.00
Reducibility (ISO 4695)	[%/min]	0.80	1.40
Drum strength +6.3 mm	[%]	65.00	78.00
Abrasion index –0.5 mm	[%]	4.20	9.80

The current state-of-the-art of the blast furnace process requires comprehensive knowledge of the quality of charge materials. For this reason, performance evaluation of Fe agglomerate currently occupies an important place in the quality control system of blast furnace charge materials. The results of a comprehensive evaluation of Fe agglomerate can be used both to manage the technology of blast furnace process, but also to manage the technology of agglomerate production itself. The final properties of the agglomerate depend on numerous factors (about 30) that affect its quality. This is a rather complex issue, and the aim of this publication does not include discussing these factors. However, one of the crucial factors affecting the final properties of the agglomerate and the economic and ecological conditions of its production – the effect of carbonaceous fuel – will be described later.

7.1.1. Requirements related to carbon fuel for agglomeration process

In the agglomeration process, bituminous coal coke, grain size < 3 mm, is used as a fuel. It is actually undersize blast furnace coke with a grain size of 0–10 mm, which is milled to a grain size below 3 mm. Such modified carbonaceous fuel is called agglomeration coke (or coke powder), *Fig. 48*. In addition to traditional coke powder, anthracite, lignite coke, coke of slightly sintering coal, and lignite can be used as a substitute fuel.

From the chemical point of view, these properties are the criteria of coke powder quality:

- Non-volatile carbon content (C_{FIX}) – as a basic constituent of coke organic matter, it occurs in the range 83–91 % and determines the calorific value of coke.

- Volatile substances (VM) – mainly CO, H_2, CH_4 and CO_2. Their content reflects the level of coal charge coking or degree of coke quality. Volatile substance content depends mainly on the coal charge heating, carbonisation temperature and time. Coke powder has a volatile matter content of about 2.0 to 4.0 %.

- Ash content (A) – ranging from about 8 to 10 % in top quality types of coke powder. Standard coke powder generally has a higher ash content – approximately 10–14 %. The standard analyses of the ash usually show 40–55 % of SiO_2, 5–30 % of Al_2O_3, 15–30 % of Fe_2O_3, 1–5 % of CaO, 0.5–3.0 % of MgO and 0.2–2.0 % of P_2O_5.

- Sulfur – it is for the most part (65–80 %) organic and the rest of the sulfur is in the form of FeS_2, $FeSO_4$, $CaSO_4$ and the elemental sulfur. Its content in the coke powder is approximately from 0.62 to 2.5 %. The organically bound sulfur has a strong bond, which is destroyed only at temperatures above 1500 °C.

- Phosphorus – it is an undesirable element and its content in coke depends on the content in the coal charge. The phosphorus content of the coke powder is between 0.006 and 0.283 %.

Agglomeration fuel is obtained by treatment of BF coke, so it is important to know its physical, physicochemical and thermo-technical properties:

- Moisture – water penetrates into coke during cooling or quenching. Due to the rapid cooling of hot coke, volume shrinkage occurs, and thus water or water vapour is sucked into the pores of coke. The water content of coke is increased with the reduction of grain size due to the larger surface of the relatively small grains. Its range in the coke powder is wide, from approximately 6 to 18 %.

- Actual specific density – it is determined by pycnometry and its values for coke powder are within the interval of 1400–1800 kg.m^{-3}.

- Bulk specific density – it is actually a ratio of the mass of coke in a given volume. For coke powder, its values are within the interval from 600 to 800 $kg.m^{-3}$.

- Porosity – it is influenced by the sinterability of the starting mixture, volatile matter content in coal, the degree of milling, the bulk specific weight of the mixture, and heating rate of carbonisation. Most of the pores are open and extend to the surface. To us, mainly micropores are of interest, because the coke powder has a grain size below 3 mm. With the growing number of micropores, the surface area of coke increases, which particularly affects the flammability and reactivity of coke.

- Ignition temperature – it is within the range of 550–650 °C and its value rises with the decrease of hydrogen content in the coke mass and the degree of graphitisation.

- Calorific value (CV) – it indicates the amount of heat released by full combustion of 1 kg of coke powder. It is within the range of 24–30 $MJ.kg^{-1}$.

Fig. 48 Coke powder

Flammability is also the essential property of agglomeration fuels. The combustion zone of sinter layer is moving as fast as fuel burns. Flammability depends the most on the specific surface (porosity and grain size). Other properties such as specific heat, thermal conductivity, reactivity, mechanical strength, abrasion, tensile strength, lumpiness, lumpiness deviation, shape of lumps, specific electrical resistance, and degree of graphitization are properties determined and used for lumpy BF coke or for pea coke and nut coke in the production of ferroalloys in the EAF. In the analysis of agglomeration coke powder, these are not standardly determined.

In **Tab. 16**, chemical composition and calorific values of the selected carbonaceous fuels (coke powders) for the Fe agglomerate production are given.

Tab. 16 Analysis of selected sinter fuels – coke powder for the production of Fe agglomerates

Coke powder	H$_2$O (W) [%]	Ash (A) [%]	Volatile matter (VM) [%]	Fixed carbon (C $_{FIX}$) [%]	Sulfur (S) [%]	Calorific value CV [MJ.kg^{-1}]
Slovakia	6.5	13.10	2.50	84.40	0.65	28.00
Czech Republic	5.8	12.90	2.60	84.20	0.60	28.20
Ukraine	8.2	15.10	3.50	81.30	0.70	27.20
Russia	6.9	14.15	3.25	82.60	0.72	27.95
Hungary	6.8	13.50	3.10	83.40	0.60	27.80
France	3.5	12.80	1.95	85.10	0.40	28.30
Japan	5.5	12.10	1.50	86.40	0.60	29.40
China 1	7.5	18.20	5.80	76.00	0.50	26.84
China 2	5.5	14.25	2.75	83.00	0.55	27.92
USA	5.0	12.55	2.50	85.00	0.65	28.10

Note: The analysis of ash, volatile combustible, fixed carbon and sulfur was carried out on a dry sample.

7.1.2. Use of fuel within the agglomeration process and combustion of carbonaceous fuel in the sintered layer

As stated in previous chapters, carbonaceous fuel is added in the pre-pelletisation stage of final agglomeration charge to provide the heat source for the agglomeration process. The task of pre-pelletisation is to create micropellets that increase the permeability of charge on the agglomeration belt, as well as even distribution of fuel in agglocharge. The structure of pellets can be monocentric (one component of pelletised material) or polycentric (more components of pelletised material) in the case of agglomeration charge. Polycentric structure is prevalent for the most part. In the formation of such micropellets, the grains are made of ferriferous particles, carbonaceous fuel and basic supplements. Most commonly, the core of the pellet consists of coarser ore grain or coke grain. In the case of the higher content of fine-grained concentrates in the mixture, the core may be formed by a cluster of fine grains of concentrate. Micropellets formation takes place in the presence of wetting liquid – water. The following *Fig. 49* shows the mechanism of formation of these micropellets.

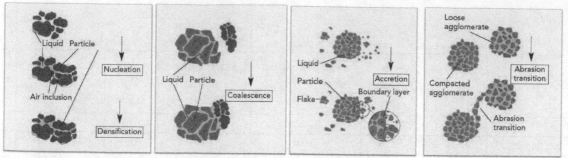

Fig. 49 Formation of micropellets

Coke particles are narrow-ringed by incombustible materials (Fe ore, Fe concentrate, limestone and lime) in the sintered layer. Hence, this process can be considered a burning of separated, isolated fuel particles in the gas stream. In the gas stream, combustion products with continuously varying temperature and chemical composition are formed.

For the sake of sintering and ensuring the combustion of carbon, it is necessary to ignite the surface of the agglomeration charge. Coke powder, located in pellets, has the ignition temperature of approximately 600 °C. Combustion of the gaseous medium in the ignition burner head generates the heat required to ignite the fuel coke, which is in the layer of pre-pelletised agglocharge. Flame temperature reaches much higher values (above 1000 °C) than the ignition temperatures of used carbonaceous fuels in the agglomeration charge. The burning of fuel never takes place in the whole agglomeration layer but runs as an ongoing process of burning in a small sintered layer (i.e. the combustion zone). Heat in each elementary layer of the charge during sintering is given by the exothermic effect of burning fuel. Due to the progress of combustion zone and major heat transfer by convection, heat accumulation occurs in elementary layers towards sintering grate. The record of static agglomeration layer sintering in *Fig. 50, 51* also proves the accumulation of heat during combustion. After the ignition of agglocharge, the combustion zone gradually moved downwards to the lower layer of the charge, increasing the maximum temperatures in the sintered layer, *Fig. 50*.

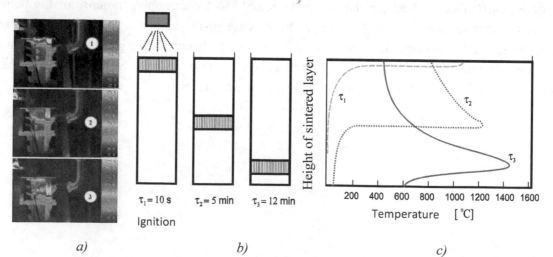

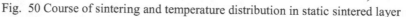

Fig. 50 Course of sintering and temperature distribution in static sintered layer

a) infrared sensing of static sintering equipment surface, b) progress of combustion zone after ignition of the charge surface, c) temperatures in sintered layer

Fig. 51 shows a visual representation of burning fuel and progress of the combustion zone in a static sintered layer. Apparently, the combustion process in the sintered layer is extremely non-homogeneous and different heat zones are formed. The representation of temperature in the sintered layer (*Fig. 50c*) demonstrates that the lowest temperature is achieved in the top layer and the highest in the bottom layer. As a result of the drying and preheating of the lower layers by off-gases from upper layers and due to the accumulation of heat in the sintered layer, the temperatures are the highest just above the grate. Non-uniform distribution of heat in the height of the sintered layers and non-uniform conditions of agglomerate formation can be eliminated by changing the distribution of fuel along the height of the sintered layer.

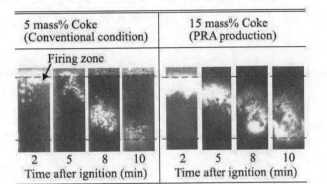

Fig. 51 Illustration of fuel combustion in sintered layer [40]

The reaction of carbon combustion takes place in the sintered layer at low (about 850 °C) and high temperatures (above 1000 °C) and has different effects on the agglomeration process. The reaction proceeds in a kinetic mode at low temperatures, and in a diffusion mode at high temperatures. Each of these modes is characterised by its own relations.

When the process operates in the kinetic mode, the resulting speed is low. In these conditions, the determining factors are the following chemical kinetics factors:

- temperature,

- physicochemical properties of the carbon grain surface,

- geometric and structural activity of the fuel grains.

The slower carbon burning rate in the kinetic mode is mainly caused by the reaction mechanism of oxidation reaction, which is slow adsorption of oxygen molecules on the surface of the carbon lattice. The reaction takes place also inside the porous fuel. When the process operates in the diffusion mode, the role of the factors listed above is reversed. As the chemical reaction proceeds rapidly and is limited by the rate of supply of oxygen molecules to the surface of the carbon grains, the speed of the gas stream and the oxygen concentration therein have the determining influence on the overall speed of the process of burning. The influence of temperature and physical-chemical properties of the surface of the carbon grains is smaller. Unlike the kinetic mode, where the reaction may be carried out within the porous fuel, in the diffusion combustion mode, the combustion takes place only on the outer surface of the fuel particles.

Now, let us take a closer look at the processes related to the burning of fuel (coke powder) and affect the physical-chemical and thermal processes taking place in the sintered layer.

The process of immediate combustion of carbon fuel is preceded by water evaporation and thermal decomposition of the fuel, which undergoes the separation of the volatiles in the temperature range of 300–900 °C (a major portion is vaporised within the temperature range of 300–600 °C), *Fig. 52*. The individual processes in *Fig. 52* are shown within a small sintered layer – 10 cm (i.e. combustion zone) – and the subsequent cooling sucked air. The Total actual sintering time throughout the agglomeration layer (40 cm) is normally 20 to 30 minutes.

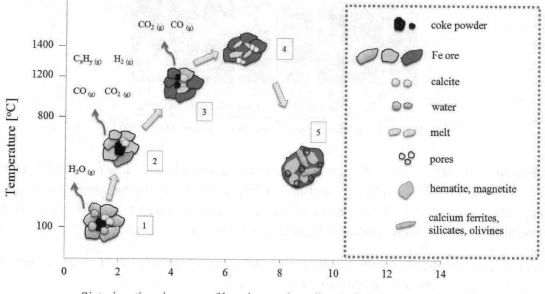

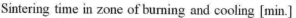

Fig. 52 Illustration of processes associated with fuel combustion and formation of agglomerate structures during sintering

(1) – evaporation of water, (2) – thermal decomposition of fuel and dissociation of carbonates, (3) – fuel combustion and completion of carbonate dissociation, beginning of grain melting, (4) – intense heating, formation of melt and forming of primary structure of agglomerate, (5) – agglomerate cooling, formation of resulting porous structure of agglomerate

The actual mechanism of combustion of carbon is as follows:

- adsorption of oxygen molecules on the surface of the carbon lattice,

- mutual chemical action of oxygen and carbon with the formation of C_XO_Y complexes of indeterminate composition,

- disintegration of C_XO_Y complexes to CO_2 and CO, and their desorption into the outside atmosphere.

According to this mechanism, primary products of combustion are simultaneously CO and CO_2. This process can be noted in the form of two parallel reactions (7 and 8):

$$C + O_{2\,(g)} = CO_{2\,(g)} \qquad \Delta H_{298} = -395 \text{ kJ.mol}^{-1} \qquad (7)$$

$$C + 1/2\ O_{2\,(g)} = CO_{(g)} \qquad \Delta H_{298} = -113 \text{ kJ.mol}^{-1} \qquad (8)$$

The CO/CO_2 ratio in combustion products is significantly affected by the basic state parameters – temperature and pressure. Since both reactions are exothermic, raising temperature shifts the equilibrium to the left. However, this balance shift is more significant for reaction (7) than for reaction (8), whose heat effect is several times smaller. Therefore, the temperature increase will create more favourable conditions for reaction (8), e.g. the CO/CO_2 ratio in the combustion products increases. In contrast to the temperature, increasing gas pressure deteriorates the thermodynamic conditions of reaction for the combustion to CO, since the reaction (8) runs with increasing volume of gaseous products. The CO/CO_2 ratio drops with the increasing pressure.

Since the combustion of the solid fuel is a heterogeneous process, running between the solid (coal) and gas (oxygen in the atmosphere) phase, the speed of its course also depends on the feed rate of oxygen to the reaction surface (by diffusion) and the rate of CO_2 and CO formation.

During sintering, the oxygen content always decreases and the content of CO and CO_2 in the combustion products increases. By the end of sintering, when the combustion zone shifts to the lower layers, the process of gaseous component removal is reversed – the oxygen content increases and the content of CO and CO_2 in the combustion products decreases. The sintering process is finished, when the oxygen content in the exhaust gas is about 20.5 to 21 %, and the content of CO and CO_2 is close to zero, *Fig. 53*.

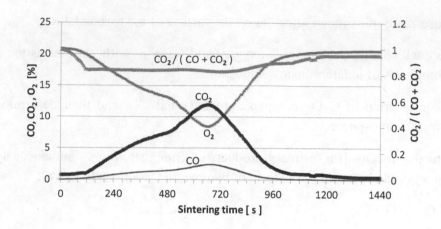

Fig. 53 Illustration of gaseous component content during sintering

Not all oxygen reacts with carbon fuel due to the isolation of coke particles. The combustion of fuel in the layer, through which air is sucked, is a considerably more complex process. Under these conditions, in addition to reactions (7) and (8), the following reactions (9 and 10) may take place:

$$C + CO_{2\,(g)} = 2\,CO_{(g)} \qquad \Delta H_{298} = +167\,\text{kJ.mol}^{-1} \qquad (9)$$

$$CO_{(g)} + 1/2\,O_{2\,(g)} = CO_{2\,(g)} \qquad \Delta H_{298} = -281\,\text{kJ.mol}^{-1} \qquad (10)$$

Another reaction runs in the sintered layer as well, which has a highly exothermic effect – the reaction of oxidation of magnetite to hematite (11):

$$4\,Fe_3O_4 + O_{2\,(g)} = 6\,Fe_2O_3 \qquad \Delta H_{298} = -479\,\text{kJ.mol}^{-1} \qquad (11)$$

Despite the fact that excess air is always present in the sintering atmosphere, the combustion is never complete, i.e. there is always a certain amount of CO present in the exhaust gas. The presence of CO can be explained by the rapid flow of air that carries away CO produced by burning fuel to a cold zone of the layer so rapidly that the oxidation by atmospheric oxygen does not proceed completely. This explanation is supported by the fact that the increased negative pressure causes increased linear flow rate through the sintered layer, resulting in increase in the CO/(CO + CO$_2$) ratio.

Amount of produced CO is mainly related to the amount of solid fuel in the sintered layer. At higher fuel content, there is further decrease in the CO$_2$/CO ratio due to higher temperatures. However, the total concentration of CO$_2$ and CO in the exhaust gas is increasing proportionally with the increase in fuel consumption.

7.1.3. Processes occurring in the sintered layer

Due to high temperatures generated by the combustion of carbonaceous fuel, complex chemical and physical processes occur in the sintered charge. In these processes, the surface of the particles is reduced, decreasing the surface free energy of the system. Therefore, sintering is an irreversible process, which determines the mechanical and metallurgical properties of the sintering product. *Fig. 52* schematically illustrates the basic process relating to the fuel combustion and forming of the agglomerate structure during sintering.

In the sintered layer, several zones are formed, which differ according to the nature of the running processes. Given the production of the agglomerate requires heat, carbon combustion, as described in the previous section, is the most important reaction in the sintered layer. Thermal processes in the sintered layer affect the course of chemical reactions and the resulting structure of produced Fe agglomerate. Combustion of carbon in agglomeration charge is accompanied by a range of other simultaneous reactions, *Fig. 54*, such as the dissociation of carbonates and oxides of iron, oxidation and reduction of iron oxide, which substantially changes the composition of the gas phase [41, 49].

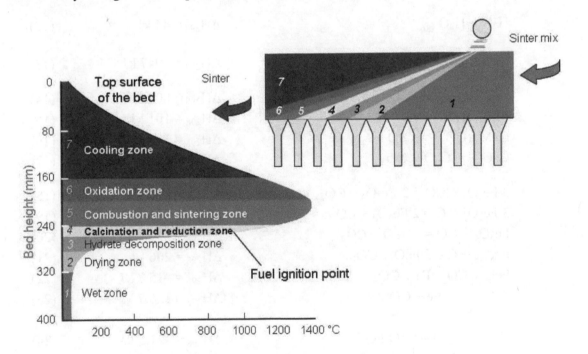

Fig. 54 Illustration of reaction zones in sintered layer [41, 49]

The chemical reactions run in individual reaction zones, *Fig. 54*. The evaporation of free water is carried out in zone 2 (reaction 12). The disintegration of crystalline and hydrate water occurs in zones 2–3 (reaction 13). Dissociation of carbonates takes place partly in zone 3, but mainly in the zone 4 (reactions 14–17). With the increase of temperature, dissociation of carbonates gradually takes place in the direction $FeCO_3$, $MgCO_3$, $CaMg(CO_3)_2$ and $CaCO_3$. Since zone 4 reaches a temperature up to about 1100–1200 °C, reduction of iron oxides may be observed in this horizon of sintered layer – either directly with carbon, or by means of the gas phase in the presence of carbon monoxide ($CO_{(g)}$) (reactions 18–23). Start of fuel combustion begins already in zone 4 when the conditions of the ignition temperature of the fuel and the presence of oxygen are achieved (reactions 7–8). In zone 5, intense combustion of fuel takes place, creating the hottest zone of the reaction sintered layer. With an increase in temperature, the more favourable conditions for reaction (8) are created, e.g. the CO/CO_2 ratio in the combustion products is increasing. Hence, the zones 5 and 6 are characterised as horizons, where the oxidation-reduction reactions take place. In these zones, the structure of the agglomerate undergoes significant changes, because in addition to reactions at the solid phase-gas phase interface, the reactions also take place at the interface of the solid phase-liquid phase (reactions 24–28). In zone 7, the cooling of agglomerate by sucked air occurs, and a final chemical reaction – oxidation of magnetite to hematite (process 11) – proceeds there.

$$H_2O = H_2O_{(g)} \qquad\qquad \Delta H_{298} = 44 \text{ kJ} \qquad (12)$$

$$Fe_2O_3. H_2O = Fe_2O_3 + H_2O_{(g)} \qquad\qquad \Delta H_{298} = -947 \text{ kJ} \qquad (13)$$

$$FeCO_3 = FeO + CO_2 \qquad\qquad \Delta H_{298} = 80 \text{ kJ} \qquad (14)$$
$$MgCO_3 = MgO + CO_2 \qquad\qquad \Delta H_{298} = 101 \text{ kJ} \qquad (15)$$
$$CaMg(CO_3)_2 = CaO + MgO + 2 CO_2 \qquad\qquad \Delta H_{298} = 303 \text{ kJ} \qquad (16)$$
$$CaCO_3 = CaO + CO_2 \qquad\qquad \Delta H_{298} = 178 \text{ kJ} \qquad (17)$$

$$3 Fe_2O_3 + CO = 2 Fe_3O_4 + CO_2 \qquad\qquad \Delta H_{298} = -50.7 \text{ kJ} \qquad (18)$$
$$3 Fe_2O_3 + C = 2 Fe_3O_4 + CO \qquad\qquad \Delta H_{298} = 121.7 \text{ kJ} \qquad (19)$$
$$Fe_3O_4 + CO = 3 FeO + CO_2 \qquad\qquad \Delta H_{298} = 33.6 \text{ kJ} \qquad (20)$$
$$Fe_3O_4 + C = 3 FeO + CO \qquad\qquad \Delta H_{298} = 206 \text{ kJ} \qquad (21)$$
$$FeO + CO = Fe + CO_2 \qquad\qquad \Delta H_{298} = -15.7 \text{ kJ} \qquad (22)$$
$$FeO + C = Fe + CO \qquad\qquad \Delta H_{298} = 156.7 \text{ kJ} \qquad (23)$$

$$CaO + Fe_2O_3 = CaO.Fe_2O_3 \qquad\qquad \Delta H_{298} = -21.5 \text{ kJ} \qquad (24)$$
$$2 CaO + 2Fe_2O_3 = 2 CaO.Fe_2O_3 \qquad\qquad \Delta H_{298} = -43 \text{ kJ} \qquad (25)$$
$$CaO + SiO_2 = CaO.SiO_2 \qquad\qquad \Delta H_{298} = -89.2 \text{ kJ} \qquad (26)$$
$$2 CaO + SiO_2 = 2 CaO.SiO_2 \qquad\qquad \Delta H_{298} = -134.5 \text{ kJ} \qquad (27)$$
$$2 FeO + SiO_2 = Fe_2SiO_4 \qquad\qquad \Delta H_{298} = -22.6 \text{ kJ} \qquad (28)$$

Due to the large surface area of the sinter charge, the use of the heat in the sintered layer is very effective. Similarly, the heat transfer contributes to the high efficiency of the process. Heat transfer in the sintered layer is executed mainly by convection, and to a lesser extent by the conduction and radiation. In each elementary layer, the heating of the charge is provided by the heat generated by:

- burning fuel in the current elementary layer,

- transfer of heat from the above layer (layers).

The heat transfer affects the heat-temperature conditions of the sintering process, and thus the completeness of the course of the chemical-mineralogical transformations and the properties of the agglomerate.

The final agglomerate contains mainly hematite and magnetite, together with calcium ferrites and silicates of different chemical composition and different morphology. Calcium ferrites generally contain certain proportions of silicon and aluminium. Thus, they are referred to as ferrocalcium silicates or silicoferrites of calcium and aluminium (SFCA). Properties of input ferriferous and basic raw materials are the crucial factors indeed that affect the final quality of agglomerate. However, the properties and the amount of carbonaceous fuel (coke powder) and high-temperature sintering technology are equally important, *Fig. 55* [36]. Chemical composition and particle size in particular are essential for the progress of physicochemical and thermal processes taking place in the sintered layer. Fuel affects not only the thermodynamics and kinetics of chemical reactions but also the amount and chemical composition of the products of these reactions. Fuel content of the agglomerate mixture determines the temperature and oxidation-reduction conditions in the sintered layer, and thereby the properties of the agglomerate.

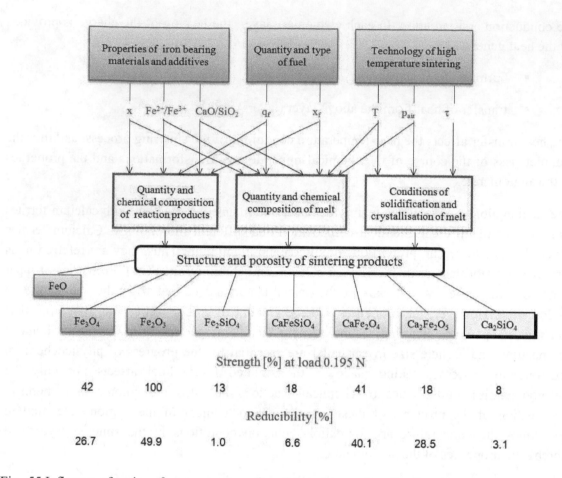

Fig. 55 Influence of various factors on properties of sintering product (Fe sinter)
[modified by authors according to 36]

(x, $_{xf}$ – chemical composition, Fe^{2+}/Fe^{3+} – proportion of iron oxide, CaO/SiO_2 – basicity
q_f – amount of fuel, T – temperature, p_{air} – underpressure of sucked air, τ – time)

The significant impact of fuel on the agglomeration process and the final quality of the produced sinter can be explained by a simple example. Carbon from fuel is necessary for the course of the agglomeration process. If carbon is low (e.g. its value does not comply with the agglomeration threshold), there is an insufficient amount of heat, and the low temperature level in sintered layer does not allow melt formation. The melt is essential for ensuring the production of so-called centres of sintering or fused grains in the sintered agglomixture, **Fig. 56**. Since the agglomeration process takes place at peak temperatures of about 1380 °C, the formation of a sufficient amount of melt with the appropriate chemical composition can be affected by changes in the composition of the agglomeration charge. In addition to the increase of fuel content, it is also possible to change the proportions of ferriferous raw materials (ore and concentrate) and the content of basic ingredients. This will create a complex Ca, Si, Fe and Al oxide-based compounds with a melting point within the temperature range achieved during sintering, *Fig. 56b*. A significant decrease in melting points of complex compounds is achieved by the presence of alkalis (Na_2O and K_2O oxides) in the charge. These are not only a natural part of iron-bearing and alkaline raw materials but are also found in the ash of carbonaceous fuels. *Fig. 56d* shows a melt, which consists of complex compounds of Si, Fe, Al, Na, and K oxides [42]. The issue of liquid phase formation in oxide systems is very well analysed by means of ternary diagrams.

Fig. 56 Microstructure of agglomerate sintering centres [42]

Varying amount of fuel in the agglomeration mixture is changing temperature conditions in the sintered layer, the amount and chemical composition of the melt, and mineralogical (phase) composition of the agglomerate. The resulting structure of the agglomerate contains not only iron oxides (Fe_2O_3, Fe_3O_4 and FeO), dicalcium silicate Ca_2SiO_4, but also the complex compounds of iron oxides with basic and acidic oxides (e.g. fayalite Fe_2SiO_4, calcium ferrites $CaFe_2O_4$, $Ca_2Fe_2O_5$, etc.), *Fig. 55*. The structure and porosity of the agglomerate have a significant effect on its mechanical and metallurgical properties. It follows from *Fig. 55* that hematite (Fe_2O_3) has the highest strength in the resulting structure of agglomerate, while dicalcium silicate (Ca_2SiO_4) has the lowest strength. Hematite has the highest reducibility. On the other hand, fayalite (Fe_2SiO_4) represents the phase that is the most difficult to reduce. Microscopic images in *Fig. 57* show the microstructure of the resulting Fe sinter. In addition to the majority grains of hematite and magnetite, there are also, e.g., ferrocalcium olivines (SFCA). These are complex compounds of SiO_2, Fe_2O_3, CaO and Al_2O_3 oxides [43].

Fig. 57 Microstructure of Fe agglomerate [43]

Hem – hematite, Mt – magnetite, SFCA – ferrocalcium olivines

A comprehensive scheme of formation of the individual compounds which are present in the resulting structure of the Fe agglomerates is shown in **Fig. 58** [36]. Obviously, the activity of ions of iron (Fe^{2+}, Fe^{3+}), calcium (Ca^{2+}), silicon (SiO^{4-}) and oxygen (O^{2-}) will have the determining effect on the structure of the agglomerate. This activity will be significantly affected by the amount of fuel and subsequent temperature conditions in the sintered layer.

The increased activity of calcium ions (Ca^{2+}) in the sintered layer will create the thermodynamic conditions favourable for the formation of dicalcium and tricalcium silicates, and calcium ferrites.

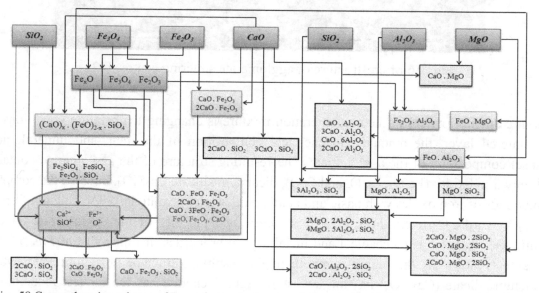

Fig. 58 Comprehensive scheme of Fe agglomerate structure (mineral phases) formation during sintering [36]

The scheme in Fig. **55** indicates that calcium silicate compounds have the lowest strength (especially dicalcium silicate – Ca_2SiO_4). It is often (not absolutely) the cause of disintegration of Fe agglomerates. Since the standardly produced agglomerates are primarily basic, the increased content of CaO in the mixture will contribute to the formation of higher amounts of calcium ferrite – which will have beneficial effects on the mechanical and metallurgical properties of the resulting Fe agglomerate.

7.1.4. Properties of fuels for agglomeration process

The temperature of the sintered layer and the speed of the combustion zone movement (i.e. the heat wave) depends on the relationship between combustion and heat transfer processes. In chapter 6, an important property of carbonaceous fuels – reactivity – was mentioned. It is also used to assess the degree of oxidation of the carbonaceous fuel. *Fig. 59* shows the morphology of the carbonaceous material before and after oxidation [44]. The oxidation of carbonaceous particles leads to a change in their morphology, which is also demonstrated by the change of light reflectance of basic macerals in carbonaceous fuel. The more reactive fuel is the one that changes more due to the temperature and oxidation, and for which the degradation of material is more significant. Reactivity tests are carried out methodically in a way that allows the observation of changes in morphology and petrographic composition of the carbonaceous fuel after the test as well. *Fig. 60* shows the microstructure of the carbonaceous material before and after the partial oxidation. Fuels with medium reactivity (in relation to the CRI index these are the values of approximately 33–40 %) are the most suitable for the agglomeration process [44].

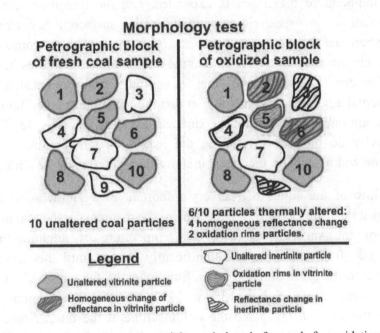

Fig. 59 Illustration of carbonaceous material morphology before and after oxidation [44]

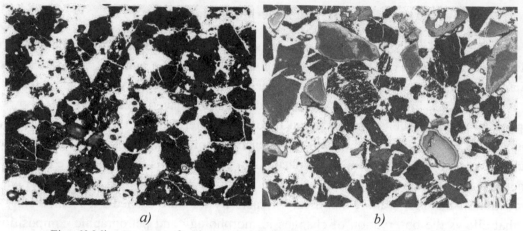

a)   *b)*

Fig. 60 Microstructure of carbonaceous fuel before (a) and after oxidation (b) [44]

The fuel with high reactivity and a large surface area will burn up on the given horizon before the heat from the upstream combustion zone reaches this horizon, *Fig. 61*. The heat from layers burning upstream is not used to heat the air participating in the burning of a given horizon. The temperature maximum is expanded, and the maximum temperature is low, *Fig. 61*. If the fuel has low reactivity and small reaction surface, it is not able to burn up on the horizon when the horizon gets the heat from the upstream combustion zone. The temperature maximum is again expanded and the maximum temperatures in the sintered layer are also low. The conclusion is that any fuel for agglomeration process should have optimum reactivity for actual agglomeration mixture so that the temperature maximum is not expanded and that the maximum temperature in the sintered layer is high, *Fig. 62*. The fuels with the extreme reactivity do not allow reaching the maximum temperatures and the minimum combustion zone width, and are, therefore, inconvenient in terms of the sintering process.

The burning time of the solid fuel is very important in agglomeration of iron ores and concentrates, as it determines the residence time of the materials at the maximum temperature. At the start of the sintering process, the temperature of agglomixture is raised to approximately 60–80 °C and does not significantly change until the approximation of the combustion zone to the given sintered layer. Subsequently, the temperature rises rapidly and reaches the maximum value of about 1380 to 1450 °C within 2 to 3 minutes. The maximum temperature of the sintered layer and the holding time at the temperature of agglomixture depends on the heat released by combustion of the solid fuel, the amount of heat accumulated in the sintered layer, and the gas-dynamic conditions of heat conduction in the sintered layer, *Fig. 61, 62*. The maximum temperature of the sintered layer is increased due to the increasing amount of accumulated and, consequently, recovered heat.

Each carbonaceous fuel has its own specifics, and combustion process in the sintered layer greatly depends also on the physical-chemical properties of the fuel. In that regard, it should be mentioned that various carbonaceous fuels are used for the manufacturing of the Fe agglomerate in the world. Alternative carbonaceous fuels from biomass will be specified in other chapters of this publication. As to the alternative fuels that have been studied in a laboratory and also industrially used, it is necessary to mention anthracite, lignite coke and semi-coke of poorly sintering coal.

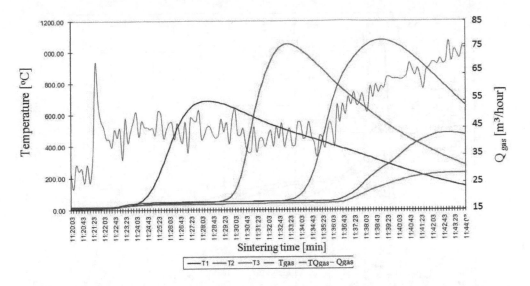

Fig. 61 Illustration of temperatures in sintered layer for fuel with extreme reactivity

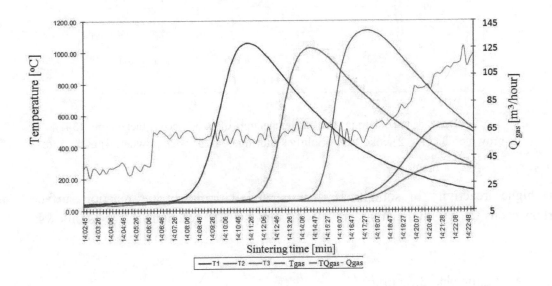

Fig. 62 Illustration of temperatures in sintered layer for fuel with optimum reactivity

The ignition temperature of brown coal semi-coke is approximately 400–450 °C. It is 600 °C for coke powder and about 650 °C for anthracite with a minimum of volatile matter. There are two types of solid fuel ignition mechanisms. In the highly active fuels containing a larger proportion of volatiles (e.g. gas coal, brown coal semi-coke), the process of release and combustion of the main part of the volatiles is followed by the process of fixed carbon burning out. In the case of fuel with lower activity (coke powder and anthracite), these two processes occur simultaneously. Combustion of individual grains of the solid carbonaceous fuel occurs at different rates, *Fig. 63* [34].

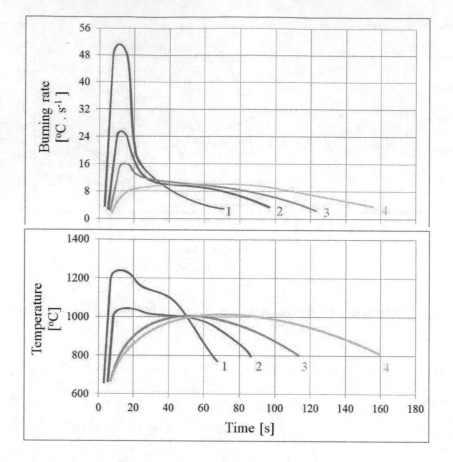

Fig. 63 Dependence of burning rate and temperature of carbonaceous fuel on time [34]
1) brown coal coke, 2) coke from slightly sintering coal, 3) coke powder, 4) anthracite

The higher reactivity of solid fuels is associated with the higher reaction surface. The carbonaceous fuels with the granularity of 1–2 mm have the reaction surface as follows:

- lignite char 3.69 $m^2.g^{-1}$,

- coke from slightly sintering coal 3.13 $m^2.g^{-1}$,

- powder coke 1.78 $m^2.g^{-1}$,

- anthracite 0.80 $m^2.g^{-1}$.

A significant temperature peak for brown coal coke, *Fig. 63*, appears to be related to the release of volatile substances (particularly hydrocarbons) and their burn up.

Reactivity of solid fuels affects the vertical sintering speed, which decreases with the reduction of solid fuel activity. The main reason is the decrease in the rate of fuel burning and deterioration of sintered layer permeability. Fuels with a high reactivity have a lower bulk density and take up more volume in the mixture, which accelerates the sintering process. Solid fuels with higher reactivity are characterised by the production of higher amounts of CO and the lower temperature of the sintered layer. For instance, when lignite coke or slightly sintering coal coke is used, the temperature of the sintered layer is standardly lower by 50–150 °C against coke powder. Anthracite has lower porosity compared to coke powder, and the thermal equivalent of anthracite has a smaller volume. Therefore, the permeability of the sintered layer is lower. The combustion process is the slowest process in the sintered layer due to the low reactivity of anthracite. It also causes a decrease in the vertical speed of sintering. Since higher temperatures are generally in the lower part of the sintered layer, it is possible to use anthracite in the lower layer in dual-layer sintering.

According to the research conducted, a mixture consisting of 50 % of coke powder (grain size 0.5 to 3.0 mm) and 50 % of anthracite (grain size 1.0 to 3.0 mm) was used in sintering [34]. The total amount of carbon in the agglomeration mixture was 5 %. In the first series of experiments, the fuel mixture was added. In the second series, anthracite dust was added to the bottom of the sintered layer. The results have shown that the sintering speed increased from 23.1 $mm.min^{-1}$ to 27.5 $mm.min^{-1}$ with the same yield and quality of the agglomerate, *Fig. 64*. An increase of the mean particle size of fuel has also increased the vertical speed of sintering. *Fig. 64* shows the maximum temperature in the combustion zone at each level of the sintered charge. The maximum temperature was higher in the bottom layer (closer to the grid) and rose due to the increase of the fuel mean particle size as well.

The use of coal in the agglomeration process is problematic too. The high content of volatile substances, which are released from the fuel at temperatures of 150–700 °C in the preheating zone, evaporates from the process with the relatively low utilisation. The oxygen, which is present in the sintered layer of the sucked air, reacted with the released volatiles (mainly CH_4 $_{(g)}$, CO $_{(g)}$, CO_2 $_{(g)}$, H_2 $_{(g)}$, H_2O $_{(g)}$ and H_2 $_{(g)}$) indeed. However, the heat effect of oxidation of these components is lower compared to the perfect combustion of carbon. The lower heat effect also contributes to the condensation of water vapour at lower levels of sintered layer. The negative factor is also the condensation of certain volatile substance release and oxidation products and their deposition in the gas line of sinter plant. For example, deposits of tar substances are thus formed, reducing the exhauster power. In addition, volatile substances increase the volume of flue gas, worsening the aerodynamic conditions and decreasing the performance of the sinter plant.

An interesting example is the sinter plant in Germany, where with the use of gas coal containing 6 to 6.5 % of ash and 35 to 41 % of volatile substances, high-quality Fe sinter was produced (comparable quality to the use of coke powder) [34]. Several hours of operation, however, caused the clogging of gas pipe of the agglomeration plant with pitch, and considerably increased losses of the negative pressure in a multi-cyclone. For this reason, and because of the possibility of explosion of electrostatic filters, the content of volatile matter in the solid carbonaceous fuel for the agglomeration was, until recently, limited to 5 %. At present, however, thanks to the technical possibilities of more advanced agglomeration plants with close monitoring of the chemical composition and quantity of flue gas, this problem can be avoided.

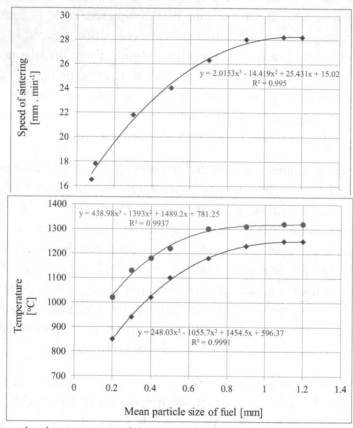

Fig. 64 Changes in speed and temperatures of sintering in combustion zone, depending on size of fuel particles [34]
1) maximum temperature at 180 mm from grate, 2) maximum temperature at 80 mm from grate

A prerequisite for obtaining high technical-economic indicators of agglomeration equipment operation is the minimal amount of fine-grained components from 0 to 0.5 mm in the used coke powder. The share of this grain size class in agglomeration fuel reaches 25–35 %. As a result, the optimal grain size of fuel mix gas-dynamic and thermodynamic conditions are improving in the sintered layer and effectivity of carbon in the fuel is increasing. A fuel with grain size of 1 to 2 mm can currently be viewed as an optimum. Opinions on the optimum granularity of agglomerate fuel have quite a long historical development. In the fifties of the 20th century, the optimal grain size of the agglomeration fuel was stabilised within the range of 0–5 mm. Gradually, the upper limit of particle size was reduced to 3 mm, and the bottom was increased to 0.5 mm. It is expected that opinions on the optimal particle size for fuel agglomeration process will continue to evolve. We should not forget that each specific composition of agglomerate charge requires its own approach regarding the optimal granulometric structure of coke powder (or any carbonaceous fuel). As said, the fine-grained fuel (with a high share of 0–0.5 mm grain sizes) is disadvantageous for the agglomeration process. Besides reducing vertical sintering speed and agglomeration equipment performance, a fine-grained fuel also causes a decline in the agglomerate quality and increases the quantity of unused carbon. The particle size of fuel significantly affects such properties of the agglomerate as strength, reducibility, a temperature range of softening and melting as well as the strength after reduction, *Fig. 65* [34].

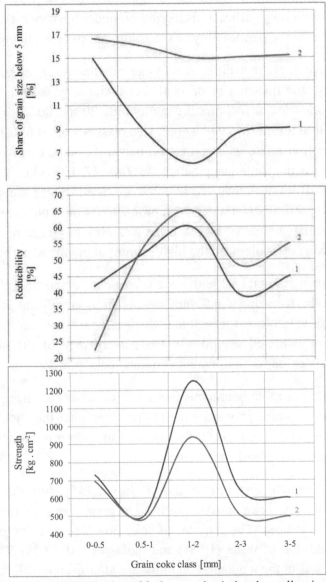

Fig. 65 Effect of particle size distribution of fuel on mechanical and metallurgical characteristics of agglomerates [34]

Legend: 1) Fe ore agglomerate, 2) Fe concentrate agglomerate

Coarse-grained fuel reduces burn rate and vertical speed of sintering. It does not burn in the combustion zone and reduces the maximum temperature in this zone. Moreover, due to the uneven distribution of carbon, the thermal inhomogeneity of the sintered layer and the number of unsintered places increases. The coarse-grained fuel creates large aggregates of sinter with large pores in the centre. In such places, there are often remains of unburned carbon. The agglomerate sintered with the coke powder above 2 mm particle size contains more dust fractions (i.e. higher proportion of the return agglomerate of less than 5 mm grain size). Also, agglomerate sintered with coke powder under 0.5 mm particle size contains a large amount of dust fractions. As expected, the highest reducibility is achieved by agglomerate produced with the fuel of 1–2 mm grain size. The fuel of 1–2 mm grain size can be considered at present the optimum for iron ore agglomeration because the agglomerate made with this fuel retains very stable characteristics with no significant disintegration under the conditions of a blast furnace, e.g. it does not deteriorate the permeability of the column of material in the blast furnace.

When using alternative fuels that differ from coke powder in chemical composition, bulk density and reactivity, the question of optimal fuel granularity is even more important. In studies conducted worldwide, a fuel mixture comprising 20 to 100 % of bituminous coal in grain size class of 0 to 0.5 mm (or 100 % grain size classes from 0 to 0.1 mm) was used for sintering of agglomeration mixtures. In all experiments, the same carbon content, i.e. 4.4 %, was used in a charge of the same composition. In *Tab. 17*, individual indicators within the agglomeration process from the conducted research are listed. It was found that in addition to reducing the vertical speed of sintering and the productivity of agglomeration plant (due to a reduced permeability of the sintered layer), there was also an attendant phenomenon of the decrease in the quality of the sinter, and the high amount of unburned carbon.

The results obtained using brown coal semi-coke of various composition showed that removal of the classes of grain size under 0.5 mm results in a significant improvement in the agglomeration process variables. If the grain size class below 0.1 mm is removed from the brown coal semi-coke (sintered at 900 °C), the performance of the sinter plant will be comparable to those observed when using conventional coke powder.

When using different types of carbonaceous fuels, it is necessary to change the upper limit of particle size distribution, which should be the higher the higher the chemical activity of the carbon in the fuel is. The requirement for the maximum proportion of small grains in agglomeration fuel, however, should not change.

Another interesting research was carried out in Egypt [45]. For the production of the agglomerate in a laboratory, coke powder of different grain size (1–3 mm, 0–3 mm and the granules of 1–3 mm) was used. The granules of coke powder grain size below 0.5 mm were pelletised, with the addition of water (10 %) and molasses (10–17 %). The granules formed were then dried and used in the agglomeration mixture after drying. Effect of adding a binder in the form of molasses was reflected in increased productivity of agglomerate production up to 13 % of molasses addition. The productivity declined at higher levels of added molasses.

Tab. 17 Indicators of sintering process using fine-grained coal shares in agglomeration mixture

Indicators of sintering	Share of granulometry class 0–0.5 mm in fuel [%]					Share of class 0–0.1 mm in fuel [100 %]
	0	20	40	60	100	
mean diameter of fuel particles [mm]	1.26	0.70	0.48	0.36	0.25	0.05
flue gas temperature [°C]	420	440	400	380	335	320
productivity of equipment [$t.m^{-2}.hour^{-1}$]	1.83	1.67	1.47	1.25	0.90	0.67
vertical speed of sintering [$mm.min^{-1}$]	28.0	26.2	24.0	22.0	20.0	16.8
Chemical composition of agglomerate						
Fe_{TOT} [%]	54.60	54.93	53.95	53.20	53.10	53.05
CaO/SiO_2	0.97	0.98	0.95	0.99	0.99	0.95
$C_{residual}$ [%]	0.001	0.001	0.014	0.019	0.047	0.10

The samples of the produced agglomerates were subjected to the test of reducibility (at 800 °C and a hydrogen flow rate of 1.5 l/min). The agglomerates, which were produced with the coke powder of the grain size class of 1–3 mm (including the granules), had better reducibility because of higher porosity. The lowest porosity was observed in the case of agglomerate produced by unseparated coke powder utilisation. The evaluation of X-ray diffraction analysis of the agglomerates showed that, because of the low temperatures in the sintered layer, the wüstite mineral phase was almost absent when unseparated coke powder was used, *Fig. 66*.

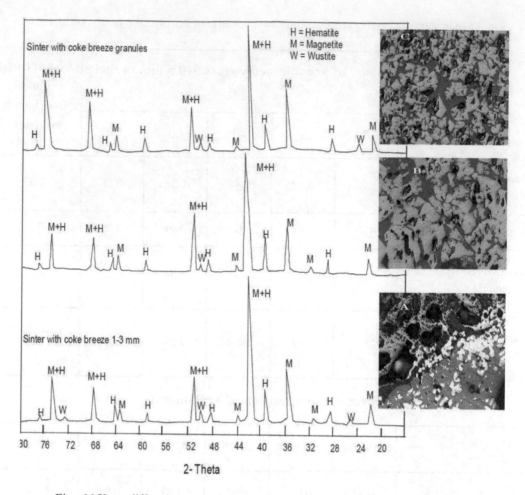

Fig. 66 X-ray diffraction analysis and microscopic structure of agglomerates [45]
a) coke powder 1–3 mm, b) coke powder 0–3 mm, c) coke powder granules 1–3 mm

7.1.5. Effect of fuel quantity on agglomeration process and the quality of Fe agglomerate

The issue of determining the required amount of carbonaceous fuel for the production of Fe agglomerates is crucial. The change of fuel quantity in agglomeration mix changes almost all technological and qualitative indicators of agglomerate production. Fuel has a critical influence on temperature and temperature oxidation-reduction conditions in agglomeration charge. Standard amounts of coke powder in agglomeration mixture are currently approximately 2.8–3.5 %. In the 80's of the 20th century, the standard fuel content in agglomeration mixture was about 4.5 to 6.5 %. Of course, this content is different due to the composition and quality of input materials, their moisture, the choice of basicity of the produced agglomerate, intensification measures, heat regeneration, etc. Fuel quantity values in the agglomeration mixture represent the average values for the sintered layer. Since carbonaceous fuel contains, besides carbon, also other components (fly ash, water, volatile substances, etc.), the total amount of carbon in the sintered layer is always lower than the fuel content. Changing fuel content in the sintered layer causes a change in the technical and economic parameters of the sintering process and change in the Fe agglomerate quality in a wide range. Therefore, it is necessary to determine the optimal fuel content for each agglomeration charge. Increasing fuel content in the agglomeration charge enhances the Fe^{3+} transition to Fe^{2+}, increases the number of liquid phases, and their degree of overheating. The more Fe^{2+} cations are created and pass into the melt, the less high iron oxide (Fe_2O_3, Fe_3O_4) remains, agglomerate strength improves, yet the reducibility decreases. FeO content is an essential indicator of Fe agglomerate quality. With increasing content of fuel from 3.5 to 5−6 %, FeO content grows from 8 % to 16 to 20 %. This creates reducing conditions for the reduction of higher iron oxides (Fe_2O_3, Fe_3O_4) to FeO or even small amounts of Fe metal. At the higher content of fuel, there are higher maximum temperatures in the sintered layer, at which reduction runs to a greater extent. ***Fig. 67 and 68*** show the morphology of the non-reduced and reduced hematite ore particles [46]. Due to the higher temperatures, it leads to an accelerated motion of individual atoms and to an increase in the interatomic distances, which result in the more intense transfer of oxygen out of the oxidic bond. The structure of the reduced particles at the higher temperature is more porous, and the pores are gradually filled with the originating melt. At the solid phase (reduced iron oxide grains) – liquid phase (melt formed by complex compounds based on oxides of Ca, Si, Fe, Al, Na and K) interface, calciumferritic olivines and fayalite are formed primarily.

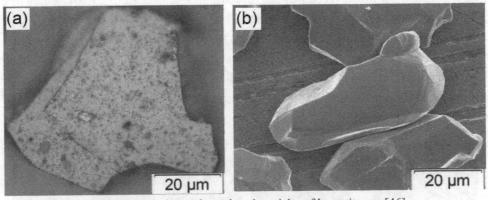

Fig. 67 Morphology of unreduced particles of hematite ore [46]
a) optical microscope. b) SEM analysis

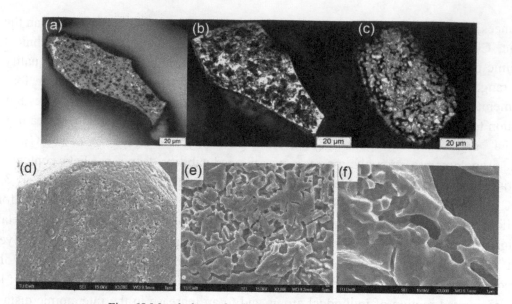

Fig. 68 Morphology of reduced hematite ore particles [46]
a) optical microscope 1277 °C, b) optical microscope 1327 °C, c) optical microscope 1377 °C, d) SEM analysis 1277 °C, e) SEM analysis 1327 °C, f) SEM analysis 1377 °C

At the low levels of the fuel in agglomeration, mixture temperature is low and melt viscosity is high. These give rise to a brittle glass in the structure of sinter and its low strength. At high fuel contents in the agglomeration mixture, the SiO_2 content in the agglomerate increases, because the burning of more coke powder creates more ash, **_Fig. 69_**, of which the major component is the very SiO_2 [47, 48]. SiO_2 is at higher contents of fuel mainly in the form of calciumferritic olivines and fayalite (Fe_2SiO_4). The very fayalite mineral phase plays a very important role in shaping the structure of the resulting Fe agglomerate. The composition of the melt in the agglomerate is in most micro-volumes in the $Fe_3O_4 - FeO - Fe_2SiO_4$ system. When the amount of wüstite is not large, the melt crystallisation may be explained by the diagram of the $Fe_3O_4 - Fe_2SiO_4$ system.

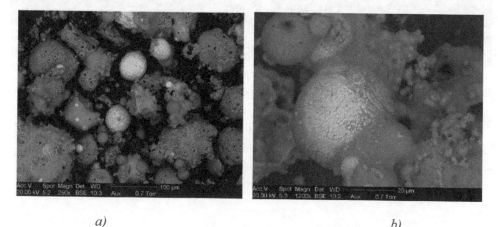

_a)_   _b)_

Fig. 69 Morphology of ash particles [47]
a) non-melted and melted grains, b) detail of melted grain

In agglomerates produced with low fuel content, primary hematite and SiO_2 appear in the structure. Magnetite is present most often at pore boundaries and cracks. In agglomerates with standard and increased amount of fuel, the bulk of the agglomeration charge passes through the liquid state, and the agglomerate structure is formed by crystallisation from the melt, *Fig. 70* [49].

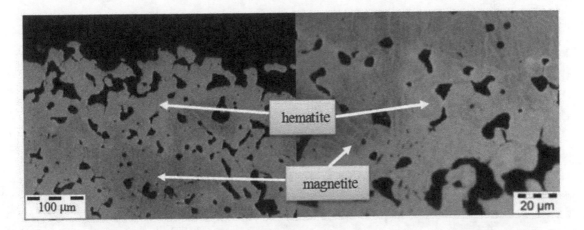

Fig. 70 Representation of hematite and magnetite in resulting Fe agglomerate structure [49]

In *Fig. 71*, there is a simplified mineralogical composition of acidic (B = 0.8) and basic (B = 1.6) Fe agglomerate at different coke powder carbon consumption [34]. At low carbon contents, the structure of acidic and basic agglomerates is mainly composed of higher iron oxides – mostly hematite. By increasing the carbon content, the reducing conditions are created, the proportion of magnetite and wüstite in the structure of the acid agglomerate increases, and the proportion of ferrocalcium olivines increases as well. With increasing carbon content, the proportion of magnetite in the structure of the basic agglomerate increases, the proportion of the calcium ferrites increases significantly, and the proportion of ferrocalcium olivines increases as well. At approximately 13 % of carbon, there is a significantly higher proportion of calcium ferrites and ferrocalcium olivines in the structure of the basic agglomerate than in the structure of the acid agglomerate.

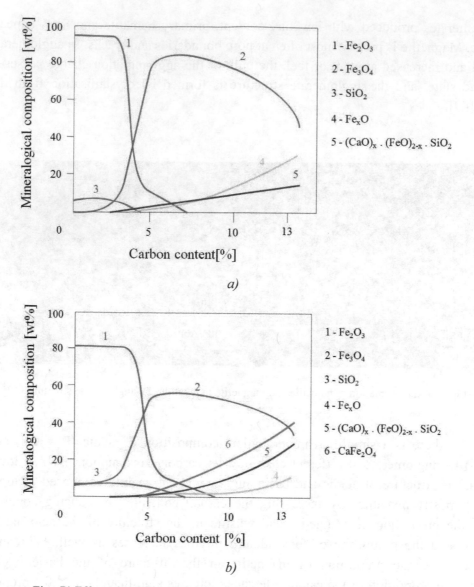

Fig. 71 Effect of carbon on mineralogical composition of Fe agglomerate [34]
a) acidic agglomerate, b) basic agglomerate

Fig. 72 illustrates the effect of the carbon content in the agglomeration mixture on individual sintering indicators and properties of the produced agglomerate.

Changing fuel content in the sintered layer results in a change in the technical and economic parameters of the sintering process and the quality of the produced agglomerate in a wide range. Therefore, it is necessary to determine the optimum fuel content for each agglomeration charge. With very low fuel contents, the resulting strength of the agglomerate is low due to the low temperatures in the sintered layer.

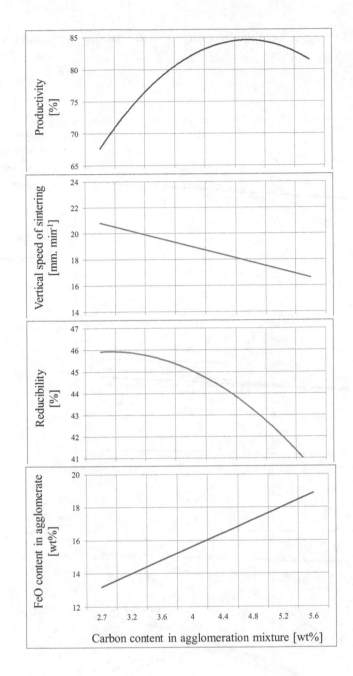

Fig. 72 Effect of carbon content in agglomeration mixture on indicators of sintering and properties of agglomerate [34]

A further increase in fuel content results in improved strength and increasing the share over 8 mm. The significant increase in fuel content, however, causes a reduction in the vertical speed of sintering. Other attendant phenomenon is an increase in the flue gas temperature, deterioration of reducibility, and increase in sulfur content in the agglomerate, *Fig. 73*. The cause of strength improvement and deterioration of reducibility of the agglomerate is the increase of FeO content and decrease of Fe_2O_3 content in the agglomerate. Content of SiO_2 and CaO in the resulting agglomerate increases relative to agglomixture, and due to the inhomogeneity of the agglomerate, there are also deviations within its various grain size classes.

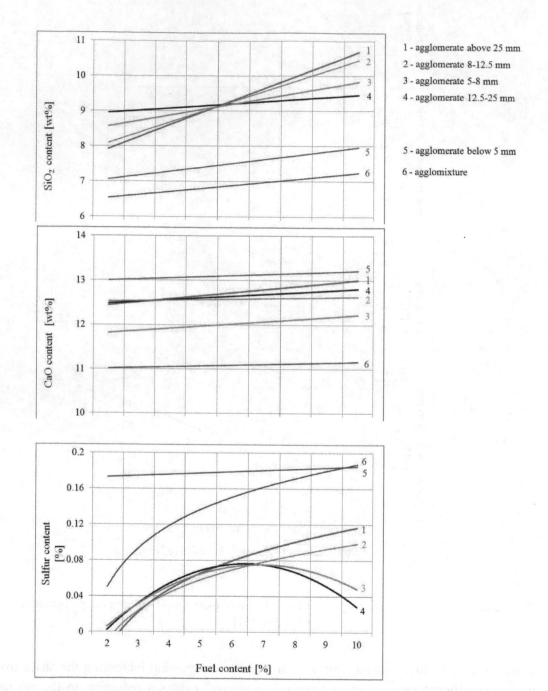

Fig. 73 Dependence of SiO_2, CaO and S content in agglomerate on fuel content in agglomixture [34]

7.1.6. Modelling of agglomerate production in laboratory conditions and the impact of fuel on the agglomeration process

Agglomerate production is always associated with energy consumption, the occurrence of particulate matter and flue gas emissions that are the main source of air pollution. As already described above, the technological, economic and environmental conditions of the agglomerate production are affected by a large number of factors which are difficult to evaluate. Therefore, individual parameters of the agglomeration process are usually examined and analysed in a laboratory, which in turn can give verifiable results for practical conclusions. The authors of this publication carried out many laboratory experiments, the results of which can be generalised and used to expand the information database on the use of carbonaceous fuels in the production of agglomerates. These experiments were carried out in a laboratory sintering pan (LSP), which sufficiently simulates the conditions on an agglomeration belt regarding the production of CO, CO_2, NO_X, SO_X, and particulate matter in the agglomeration process, as well as the quality of the agglomerate [50, 51].

The process of laboratory simulation of agglomerate production was divided into two stages – cold section and hot section. Within the cold section, preparation and modification of the individual components of a charge, its weighing, model heaps, pre-pelletisation, determination of moisture and permeability was performed. Within the hot section, heating of sintering pan, placing of pre-pelletised mixture into the laboratory sintering pan, ignition of the charge surface and the high-temperature sintering itself was executed, *Fig. 74.*

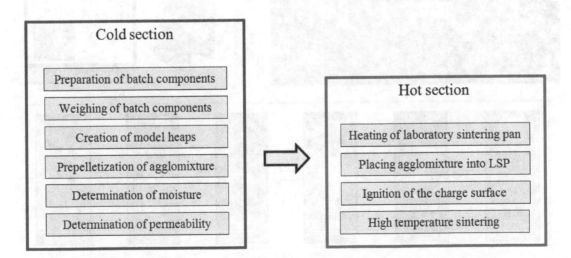

Fig. 74 Illustration of stages within simulation of laboratory agglomeration production

For the purposes of experimental sintering, a sintering apparatus – laboratory sintering pan (LSP) – was used, which is fully equipped with measuring devices and analysers, *Fig. 75.*

During sintering, fuel burning zone gradually shifted in the direction of airflow, and thus ensured the creation and cooling of the melt. The process of fuel combustion and agglomerate sintering does not occur simultaneously across the layer, but gradually in a narrow range, which is moving towards the grid. Above the zone of burning fuel, there is a segment of the finished agglomerate, which is cooled by sucked air, while part of the heat from cooling down agglomerate is transferred into the fuel combustion zone. The sintering process ends with burnout of the fuel in the bottom horizon, i.e. on the grate. In the course of sintering, various parameters were evaluated (e.g. temperature of the sintered layer, the amount and chemical composition of the flue gas, vacuum, etc.).

The temperature was measured by thermocouples. Their number and type were determined by the position in the measuring system and the measured temperature range. For the high-temperature range in the sintered layer, three thermocouples of the PtRh10-Pt type were used. The flue gas temperature was read at two levels by NiCr-Ni type thermocouple. Chemical composition and temperature of the flue gas were analysed by TESTO 350 device. The differential pressure was measured by Anubar type probe, which served for calculating the amount of sucked air (or flue gas). All quantities were read at 15-second intervals and collected in a logger. After each experiment, the collected data were transformed into a form usable on a personal computer.

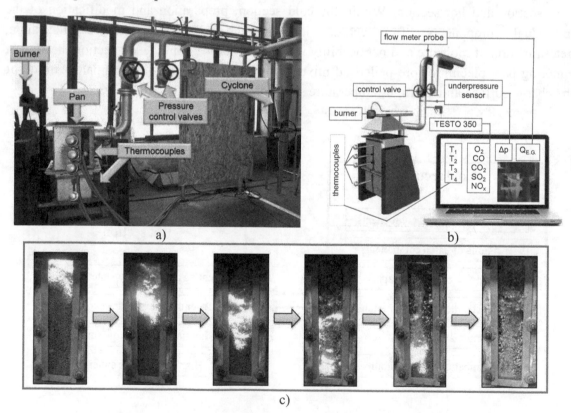

Fig. 75 Laboratory sintering pan (LSP) in Slovakia

a) technology of sintering,

b) laboratory sintering pan,

c) visible fuel burning zone

7.1.6.1. Effect of fuel quantity on CO_2 and CO content; reduction of CO_2 and CO emissions

Laboratory experiments on the laboratory sintering pan were performed using ferriferous raw materials – aggloore KRIVBAS (content of Fe_{TOT} = 60.70 %) and concentrate MICHAIL (content of Fe_{TOT} = 64.52 %). In the agglomeration process, standard coke powder with grain size < 3 mm was used as fuel (its properties are given in chapter 7.1.1). These inputs were included in prepared agglomeration mixtures with basicity within the interval of 1.7–2.8. It was, therefore, production of high-basicity agglomerates. The produced Fe agglomerates had Fe_{TOT} content within the interval of about 47–50 % at high basicity.

The theory of dependence of CO_2 and CO formation on an amount of fuel is generally known. Since the burning of solid fuel is the heterogeneous process, then its rate depends also on the rate of oxygen supply to the reaction surface and on the rate of chemical formation of CO and CO_2 itself. Quantities of CO and CO_2 emissions cannot be considered separately because they are the result of carbon fuel burning (whether complete or incomplete), oxidation-reduction, and dissociation reactions [50, 51]. In the production of high-basicity agglomerates at higher fuel consumption, CO, CO_2 and solid particulate pollutants emissions (PM) increase. The main source of these emissions is solid fuel and basic compounds based on carbonates (limestone and dolomite). Whereas in the case of increasing amount of fuel, CO and CO_2 emissions rise, the presence of basic components affects mainly the amount of CO_2 produced. By changing the different parameters of the agglomeration process, these emissions can be reduced, but the significant reduction can be only achieved by minimising fuel in agglomeration charge. The reduction in fuel content is feasible only when there is a certain heat reserve – excess heat, which can reduce the amount of fuel. The nature of the agglomeration process, i.e. the production of the agglomerate with required production, qualitative and quantitative parameters, must be maintained at the theoretically minimum amount of fuel.

In the modelling of the sintering process in the laboratory sintering pan, *Fig. 75*, the very strong dependence of the resulting amount of CO_2 and CO on the carbon content in the agglomeration mixtures was confirmed, *Fig. 76*. Laboratory experiments have provided another important insight, namely that at higher sintering speeds, the absolute amount of both CO_2 and CO was reduced, which is essential for the implementation of agglomeration technology. The rate of sintering is very closely related to the permeability of the charge. The effect of these parameters on the quantity of CO_2 is confirmed by the mathematical and statistical dependencies, *Fig. 77*.

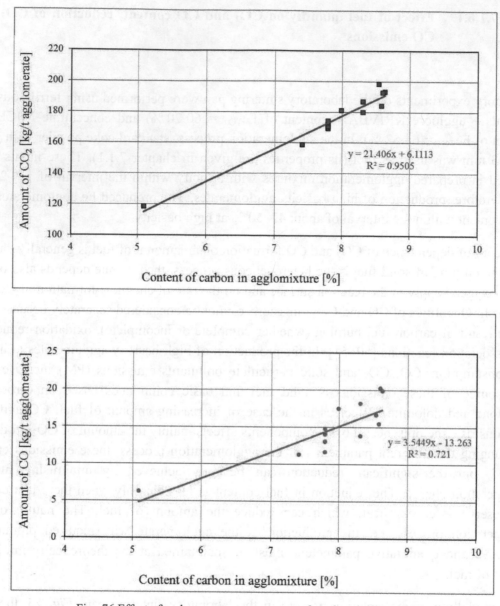

Fig. 76 Effect of carbon content on amount of produced CO_2 and CO

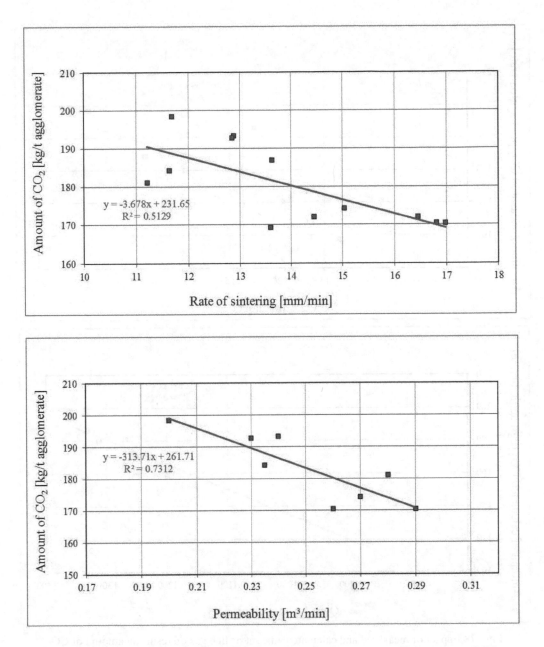

Fig. 77 Influence of sintering rate and permeability of mixture on resulting amount of CO_2

The influence on the total amount of flue gas generated in the process is also very significant. This factor was confirmed by both own measurement and theoretical calculation. Based on this, it is possible to verifiably state that there is a direct relation between the occurrence of CO_2 and the occurrence of CO, which is in direct dependence. Individual variations in the CO_2/CO ratio within the framework of the laboratory experiments involved influences of individual factors. However, it can be stated that the increased amount of flue gases in agglomeration process is reflected in the increased amount of both gaseous emissions (CO_2 and CO), *Fig. 78*.

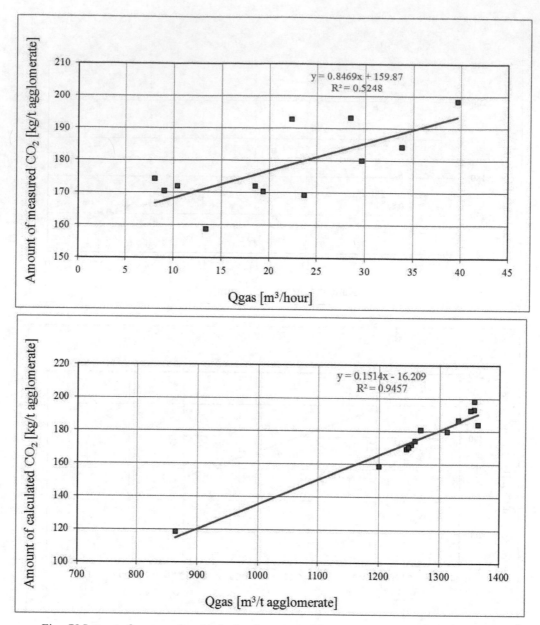

Fig. 78 Impact of measured and calculated amount of flue gas on resulting amount of CO_2

The results of laboratory experiments on the laboratory sintering pan showed that by changing or combining the settings of certain parameters of the agglomeration process, it is feasible to reduce the absolute amounts of CO, CO_2 and of particulate matter pollutants (PM) emissions. In order to achieve the required qualitative and quantitative parameters of agglomeration process with the emphasis on reducing CO, CO_2 emissions, the following measures are necessary according to the experience of the authors of this publication:

- Monitor important indicators of agglomeration charge parameters (bulk density, moisture and permeability) to optimise the charge preparation stage, and to ensure homogeneous properties of agglopellets.

- Optimize conditions of layer sintering on a strand (negative pressure, belt speed, layer height, ignition head parameters, etc.) to achieve the maximum production of agglomerate over 5 mm.

- Optimise the negative pressure, which is the main reason for exceeding the PM and gaseous CO and CO_2 emissions. Since it is not possible to measure the speed of the sucked air in the operating conditions, the speed control must be performed by opening or closing the vents in the flue-gas exhaust pipe.

- Choose the right alternative for agglomeration production. Although, at first sight, the only real possibility of increasing the agglomerate's richness is to increase the concentrate in the charge, it does not produce the expected effect. The higher proportion of concentrate degrades rollability, increases the production of PM, which is directly related to energy requirements of agglomerate production. It is the most evident in the production of high-basicity agglomerates. There is a clear dependence – low Fe_{TOT} content in agglomerate is achieved at high basicity of agglomeration charge. In addition, for the production of high-basicity agglomerates, more fuel is needed, which automatically results in higher gaseous CO and CO_2 emissions. The solution is the production of low-basicity agglomerates with a lower fuel consumption, which will automatically lead to the higher agglomerate richness and lower production of CO and CO_2.

- Selection of suitable raw materials for the production of agglomerates. With regard to the CO and CO_2 emission reduction, it is necessary to choose the raw materials, which require a lower amount of fuel for the sintering process. Ferriferous concentrates with high iron content in the form of magnetite meet this requirement. During the sintering, an exothermic reaction of oxidation of magnetite to hematite with releasing a large amount of heat takes place. Released heat corresponds to a certain amount of fuel that can be saved, thereby also reducing CO and CO_2 emissions. However, the optimal ratio between aggloore and concentrate must be chosen to ensure rollability and to achieve higher permeability of the charge.

7.1.6.2. Impact of fuel quantity on agglomerate quality

In the laboratory experiments focused on analysing the influence of fuel quantity on agglomerate quality, ferriferous input materials – aggloore KRIVBAS (Fe_{TOT} = 60.70 %) and concentrate NIŽNÁ SLANÁ (Fe_{TOT} = 53.52 %) were used for sintering. These ferriferous inputs were included in the prepared agglomeration mixtures with the basicity within the interval of 1.4–2.1. Therefore, it was the production of medium and high-basicity agglomerates. The produced Fe agglomerates had the value of Fe_{TOT} within the interval of about 46–52 %.

In the experiments, the ratio of ferriferous raw materials – aggloore/concentrate (100 % aggloore, 1:1 aggloore/concentrate and 100 % concentrate) – was changed. In this chapter, primarily the experiments with separate ferriferous raw materials are specified due to the more significant impact of fuel quantity on agglomerate quality.

Fig. 79 shows a thermal profile and Fe agglomerate made by sintering using 100 % concentrate, while coke powder was used for sintering. Due to lack of fuel, 60 kg/t of agglomerate (5.22 % of C in agglomixture), there were low temperatures of 500–850 °C in the sintered layer. The agglomerate had unacceptable properties – only some microgranules were connected. The essence of the agglomerate process – i.e. producing the agglomerate with the required production, qualitative and quantitative parameters, was not accomplished in this case. The theoretical minimal quantity of fuel is always required for agglomerate production, which depends on the physical-chemical properties of input iron bearing raw materials and alkalinity of the produced agglomerate. *Fig. 79* shows a scheme of produced agglomerate grains, which demonstrates that no melt was created at temperatures of about 800 °C.

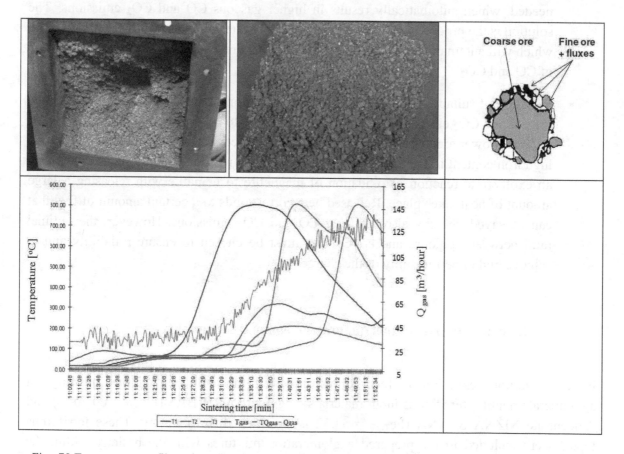

Fig. 79 Temperature profile and produced agglomerate from sintering no. 1 (100 % of aggloore, 5.6 % of C in agglomixture)

Legend: T1, T2, T3 – temperature in sintered layer, T_{gas}, T_{Qgas} – flue gas temperatures, Q_{gas} – quantity of flue gas

By increasing the amount of used fuel to 70 kg/t of agglomerate (7.5 % of C in agglomixture), standard temperatures of 1100–1250 °C were achieved in the sintered layer with 100 % of aggloore, resulting in agglomerate with the required properties, *Fig. 80*. On the surface of individual grains, the melt was formed, and due to the low viscosity of the liquid phase, multiple grains were bound together producing a sinter. Total sintering time was shorter almost by half compared to sintering no. 1 (21.42 min versus 39.33 min). The average values of CO_2 in flue gas also increased from 2.74 to 8.50 %, which confirms the already known fact that increasing the amount of fuel proportionally increases the CO_2 content in flue gas.

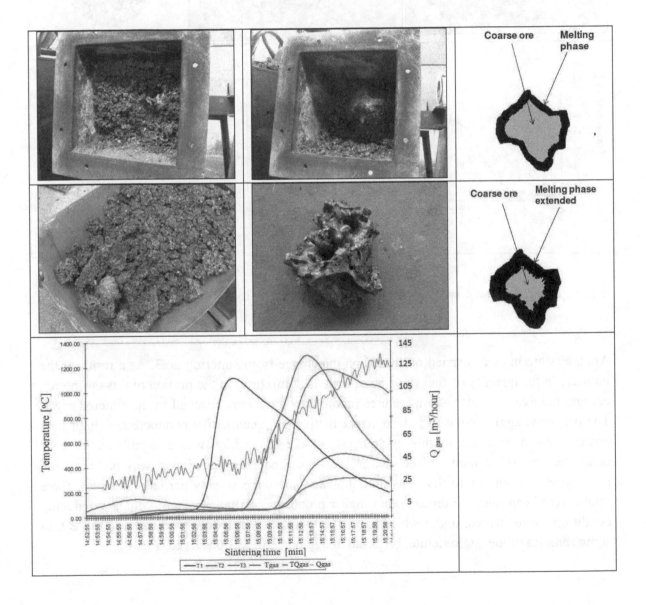

Fig. 80 Temperature profile and produced agglomerate from sintering no. 2 (100 % of aggloore, 7.5 % of C in agglomixture)

In *Fig. 81*, there is a temperature profile of produced Fe agglomerate from sintering no. 3, which was based on 100 % concentrate. Due to the insufficient amount of fuel used, 60 kg/t of agglomerate (5.22 % of C in aggloore), temperatures in sintered layer were low (500–850 °C) and the sinter had unacceptable properties. There were only several bound micro pellets.

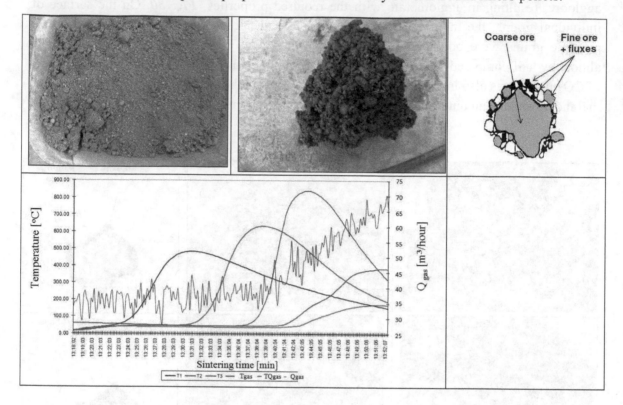

Fig. 81 Temperature profile and produced agglomerate from sintering no. 3 (100 % of concentrate, 5.22 % of C in agglomixture)

Another sintering was carried out based on the charge from sintering no.3. As a result of the increase in the quantity of fuel (6.45 % of C in agglomixture) in the portion of this unsintered charge, relatively standard temperatures (about 1200 °C) were reached in the sintered layer. The produced agglomerate had standard qualitative and quantitative parameters. A high FeO content was determined in this agglomerate (14.23 %), which was a significant increase compared to FeO content in the charge (5.2 %). Since the added fuel was probably not uniformly distributed into the individual grains (some were already partially sintered), there might have been microvolumes with a higher proportion of fuel. In these volumes, reducing conditions were created, under which higher Fe_2O_3 and Fe_3O_4 oxides were reduced to FeO. In some samples of the agglomerate, break-up of agglomerate was observed, *Fig. 82*.

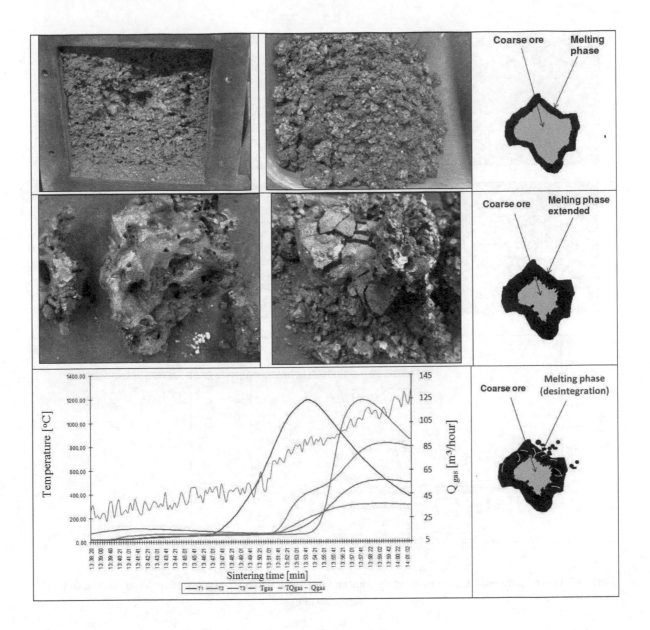

Fig. 82 Temperature profile and produced agglomerate from sintering no. 4 (100 % of concentrate, 6.45 % of C in agglomixture)

In **_Fig. 83_**, there is a temperature profile and produced agglomerate from sintering no. 5, which was based on 100 % aggloore. There was a further increase in fuel compared to sintering no. 2, when 70 kg/t of agglomerate (7.5 % of C in agglomixture) were used. At 75 kg/t of agglomerate (8.11 % of C in agglomixture), high temperatures (up to about 1300 °C) were reached in the sintered layer, resulting in agglomerate with the required properties. A part of the melt leaked into the grate.

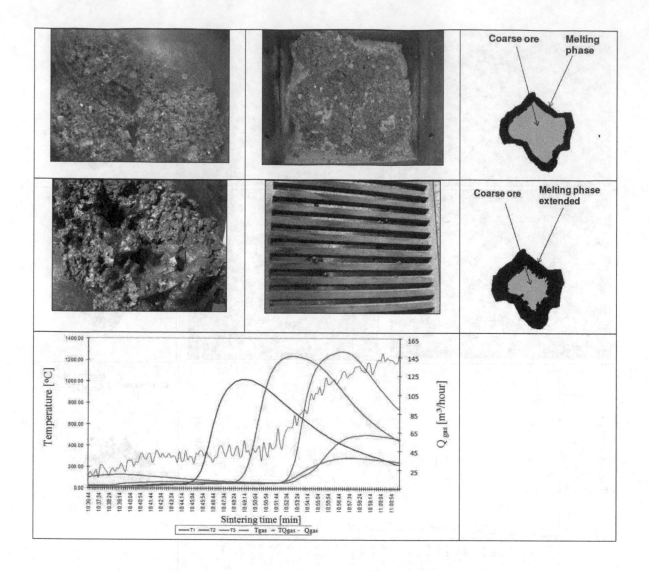

Fig. 83 Temperature profile and produced agglomerate from sintering no. 5 (100 % of aggloore, 8.11 % of C in agglomixture)

The last model sintering was carried out based on 100 % of the concentrate, *Fig. 84*. Although an increased amount of fuel was used (from 6.45 to 7.01 % of C in agglomixture), comparable temperatures (approximately 1100–1200 °C) were reached in the sintered layer as in the sintering no. 4. The produced agglomerate had standard qualitative and quantitative parameters. A relatively large portion of the melt leaked into the grate. Therefore, a large portion of the agglomerate melted onto the grate.

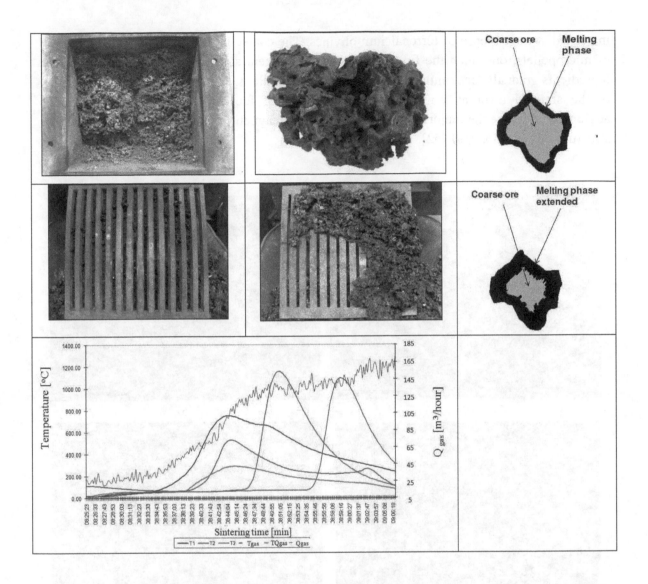

Fig. 84 Temperature profile and produced agglomerate from sintering no. 6 (100 % of concentrate, 7.01 % of C in agglomixture)

The model cases shown in *Fig. 79–84* represent only a small part of the experiments performed on the laboratory sintering pan. They were mentioned in this chapter to highlight the significant impact of the amount of coke powder on the agglomeration process. The results of laboratory experiments on the laboratory sintering pan have demonstrated that the important parameters of the agglomeration process can be significantly affected by changing the composition of the agglomeration charge (including changing the fuel content).

In relation to the change in the ratio of basic ferriferous raw materials – aggloore/concentrate – experiments were performed in the Republic of South, in which the microstructure of Fe agglomerates was analysed for a different fraction of coarse particles of Fe ore in agglomixture. *Fig. 85* shows individual model cases, from which it follows that by increasing the proportion of Fe ore coarse grains, the share of hematite in the resulting agglomerates increased at the expense of ferrocalcium olivines. The coarse grains of Fe ore form the cores of micropellets, on which the fine particles of the concentrate, carbonaceous fuel and basic ingredients gradually accumulate. Thus, created micropellet system improves the permeability of the sintered layer at a relatively lower content of fuel to form the structure of the agglomerate, where the majority phase is hematite or magnetite (darker places in the hematite structure in *Fig. 85b, c, d)* [52].

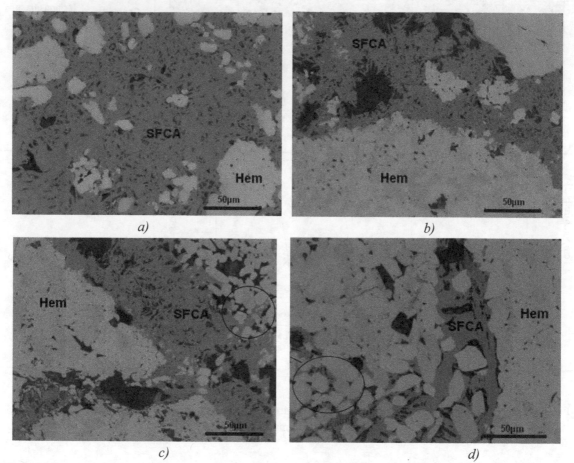

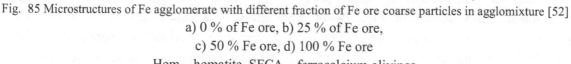

Fig. 85 Microstructures of Fe agglomerate with different fraction of Fe ore coarse particles in agglomixture [52]
a) 0 % of Fe ore, b) 25 % of Fe ore,
c) 50 % Fe ore, d) 100 % Fe ore
Hem – hematite, SFCA – ferrocalcium olivines

Pursuant to the theoretical analysis of the issue, analysis of measured values, calculations of balance, production, qualitative and quantitative parameters of the agglomeration process, it is possible to say that increasing the share of the concentrate in agglomeration charge reduces all production parameters of the produced agglomerate (yield, production and production coefficient). The main reason is the slowing down of gas dynamics due to the reduced permeability of the agglomerate. Since the course of events in the sintered layer depends mainly on the gas phase flow conditions, it is necessary to prioritise the supply of oxygen to the combustion zone. Due to the lack of air permeability of the pre-pelletised agglomeration charge based on the majority concentrate, there are no sufficient carbon combustion reactions, burning is slow (low vertical sintering rate), and temperatures in the sintered layer do not reach the required level. In such case, increasing the amount of fuel in agglomeration mixture does not help, as it is difficult to supply oxygen to the interface coke grain – gas phase (i.e. mutual chemical interaction of oxygen and carbon with the formation of the $C_X O_Y$ complexes is minimal).

7.1.7. New directions and intensifying elements in the production of Fe agglomerate

The agglomeration process is constantly improving. At present, measures facilitating the increased production and improved agglomerate quality while reducing the energy consumption and consumption of solid carbonaceous fuel are being developed and applied. These measures include:

- heat treatment of agglomerate,

- increasing the negative pressure of the air,

- agglomeration under high pressure,

- low-temperature and high-temperature heating of agglomeration charge before sintering,

- additional heating of the sintered layer by gas burners,

- sintering using flue gas from the agglomeration process.

In the heat treatment of the agglomerate to the temperature of about 1000 °C, several minutes of heating reduces the proportion of the class below 5 mm after the drum test by 10 %, the agglomerate strength increases up to 3-times, the degree of desulfurization increases by 20 %, the performance of the sintering equipment increases by 10 %, reducing the fuel consumption in the sintered layer (by 20 %). Since flue gas from the heat treatment does not contain CO, and the amount of coke powder is reduced during sintering, the CO, CO_2, NO_X and SO_X content in flue gas is lower than in the standard agglomerate production [34, 48].

The negative pressure and the suction rate are closely related and have the same effect on the sintering rate, as the suction rate increases directly proportionally to the negative air pressure. However, certain limitations and limit values for negative pressure apply to this case as well, above which the suction rate is so high that it limits the heat transfer between gas and the charge, resulting in an increase of fuel and air consumption. With further increase of the negative pressure, the charge grains can become so close and the gas dynamics resistance increase so much, that the specific air consumption will decrease again.

In the high-pressure agglomeration, compressed air is supplied from a compressor. Thereby the air pressure above the sintered layer increases several times, and hence the amount of air sucked through the sintered layer rises as well. The vertical sintering rate is markedly increased. Compaction of the charge (which can slightly reduce permeability) and activation of chemical reactions in the solid phase occur in the sintered layer. The formation of calcium ferrites and iron silicates intensifies, which leads to increasing the amount of melt and binding phase in agglomerate with the same consumption of coke powder. On the other hand, high-pressure sintering requires difficult modifications of the sintering device or the use of new equipment. High-pressure sintering enables the sintering of the charge layer up to 1000 mm thick.

External heating of the charge is performed mainly at the beginning of the sintering strand under the ignition head when the initial temperature wave is required to ignite the coke fuel in the surface layer. There is also an additional possibility of heating the charge by combined heating of the charge surface layer at about one-third of the useful length of the belt by means of gas burners burning blast furnace or other heating gas with a large excess of air. This way agglomerate quality homogenization and solid fuel savings are achieved. The negative factor is an increase of specific amounts of flue gas and the decrease of productivity of the equipment.

Preheated air, heated by cooled agglomerate in the heat exchanger to about 500 °C, has an impact on the agglomerate quality and fuel consumption, rather than a change of the temperature fields. It also has a beneficial effect on removing sulfur from the charge, increasing the degree of oxidation of the agglomerate, which also reduces the amount of air and fumes, and ensures the decline in entrained dust as well [34].

7.2. Production of Mn agglomerate

Manganese as an element is present as a component in many manganese ores in different mineralogical forms. The most stable form of manganese in nature is oxidic – either in the form of simple or complex compounds. For the needs of the industry, both oxidic and carbonate manganese minerals are useful. Although manganese ores are widely distributed around the world, the presence of rich ores is currently limited. The best manganese ores are found in South Africa, Australia, Brazil and Gabon. The Mn reserves are also found in Europe. They are mostly poorer deposits based on carbonate and complex hydrated compounds. The Mn content in these manganese ores is also very different among individual manufacturers – from about 20 to 40 %.

In the world, manganese agglomerates are produced from poorer Mn ores or from undersize particles of quality Mn ores. It is mostly the production of Mn agglomerates designed mainly for the production of FeMnC (i.e. with a low SiO_2 content of max. 8 %), and for the production of FeSiMn (i.e. with a higher SiO_2 content of about 20–30 %) to a smaller extent.

The Mn agglomerate has its justification in the charge for the production of manganese ferroalloys due to its mechanical strength and thermal stability. The use of high-quality Mn agglomerates in the process of manganese ferroalloy production does not generate a large number of small particles, which in turn affect the stability of the furnace. In the technology, the Mn agglomerate is used to increase the Mn/Fe ratio and reduce the free oxygen in the EAF. Manganese agglomerates begin to melt and reduce at the temperature that is by approximately 100 °C lower compared to the manganese ore. Due to the higher amount of difficult-to-reduce MnO, the reducibility of the agglomerate is lower, and the temperature interval of melting and reduction is higher due to the increased content of difficult-to-reduce silicates in the agglomerates. In general, there is a global search for options of optimal charge composition for the production of manganese ferroalloys, where the use of quality agglomerate is reasonable.

In **Tab. 18**, there is a chemical composition of the best manganese agglomerates in the world, which are produced from rich oxidic or carbonaceous Mn ores, and are prioritised for FeMnC production. At the same time, Tab. 18 gives the chemical composition of laboratory-produced manganese agglomerates, which are made from poor oxide, carbonate or hydrate Mn ores, and are intended primarily for the FeSiMn production.

Tab. 18 Chemical composition of manganese agglomerates [53]

Manganese sinter	Chemical composition [wt%]							
	Mn_{TOT}	Fe_{TOT}	SiO_2	Al_2O_3	P	CaO	MgO	K_2O
Comilog Sinter Gabon	58.5	3.5	7.0	6.5	0.12	0.1	0.1	0.75
Amapa Sinter JAR	49.1	9.6	7.6	7.6	0.10	0.8	0.5	0.30
Mamatwan Sinter South Africa	48.3	5.5	6.2	0.6	0.02	14.5	3.4	0.03
CVRD Sinter Brazil	54.5	4.7	5.4	8.7	0.11	1.9	0.5	1.40
Vale Sinter Brazil	57.8	2.9	6.6	6.2	0.12	0.1	0.1	0.90
Temco Sinter 75 Australia	56.0	6.5	7.1	3.7	0.11	0.4	0.1	1.50
Usinor Sinter France	54.6	6.8	6.5	8.2	0.10	2.9	-	1.10
Laboratory produced sinters from Mn ores (Turkey, Bosnia and Herzegovina, Bulgaria)								
Turkey	37.1	4.9	19.0	5.0	0.16	8.0	2.9	-
Bosnia and Herzegovina (B&H)	27.5	8.4	35.8	8.32	0.17	3.5	2.2	-
Bulgaria	36.6	6.4	23.2	4.4	0.2	5.3	6.2	-

For the production of Mn agglomerate, bituminous coal coke with grain size < 3 mm is used as a fuel. It is an undersize blast furnace coke with a granularity of 0–10 mm, which is ground to a grain size below 3 mm and is called agglomeration coke or coke powder. The properties of coke powder, which is used for the production of Mn agglomerate, are identical to the properties of the agglomeration coke, which is used to produce Fe agglomerate (section 7.1.1).

The authors of this publication also carried out experiments in laboratory conditions that focused on the production of manganese agglomerates from poor oxide, carbonate or hydrate Mn ores. The results of these experiments can be generalised and used to expand the database of information on the use of carbon fuels in the production of agglomerates [54].

In *Tab. 19*, there is a chemical composition of poor manganese ores, from which Mn agglomerates were produced. Its chemical composition of is shown in Tab. 18. It is clear that especially Mn ore Bulgaria has very high moisture (24.3 %), which limits the sintering conditions in the context of the agglomeration process. The phase analysis and thermal analysis of Mn ores showed that Mn ore Turkey is an oxidic-carbonaceous, Mn ore Bosnia and Herzegovina (B&H) is oxidic, and Mn ore Bulgaria is carbonaceous-hydrate.

Tab. 19 Chemical composition of poor manganese ores [53]

Manganese ore	Chemical composition [wt%]							Moisture [%]
	Mn $_{TOT}$	Fe $_{TOT}$	SiO_2	Al_2O_3	P	CaO	MgO	
Turkey	31.8	3.1	14.6	4.1	0.15	7.7	2.4	4.1
Bosnia and Herzegovina (B&H)	23.3	7.7	31.3	8.1	0.19	3.3	1.6	6.1
Bulgaria	29.2	3.6	18.2	3.1	0.16	4.5	3.8	24.3

Fig. 86 shows the thermoanalytical record of the sample of untreated, but homogenised Mn ore Turkey. The total weight loss in the ore represented about 16 wt%. The sample exhibited an initial moisture content of about 4 wt%, as well as a small proportion of bound water in the form of detected hydrate components of Mn, Mg, Ca, and Si oxides. The release of free water and water present in complexes corresponds to the gradual decrease of TG curves and flat stretched continuous peaks of DTG curves in a temperature range of about 90–350 °C. In the temperature range of 550–800 °C, three small temperature ranges of material release were observed, which is probably related to the presence of the detected components of rhodochrosite carbonates (29), dolomite and limestone.

$$MnCO_3 = MnO + CO_2 \qquad \Delta H_{298} = 28 \text{ kJ} \qquad (29)$$

The thermodynamic preconditions for $MnCO_3$ decomposition are realistic from 400 °C, while intense decomposition takes place only above the temperature of 550 °C. For the measured sample of Mn ore Turkey, the intense decomposition is attributable to rhodochrosite in the temperature range of 550–620 °C. The subsequent, less marked decline was apparent in the range of 620–730 °C, which may be related to the decomposition of the present ankerite. A marked drop and the DTG peak represents the release of CO_2 from the present dolomite $CaMg(CO_3)_2$ with a peak of endothermic effect at 758 °C.

Fig. 87 shows a thermoanalytical record for the sample of untreated but homogenised Mn ore B&H. In the case of the thermogravimetric record of the manganese ore sample, a linear, gradual weight loss was observed. The total weight loss represented only about 10 wt% for Mn ore B&H. The sample exhibited the input moisture content of about 6 wt%, as well as a small proportion of the detected ferrous hydroxide. The release of water corresponds to the gradual decrease in TG curves and a weak stretched peak of DTG curves in the temperature range of about 90–350 °C. The DTG record of Mn ore heating in the temperature range of 400 °C shows no significant change in the weight loss, as confirmed by the derivative of the curve, which indicates hints of two insignificant temperature intervals, namely 580–620 °C and 630–700 °C. X-ray analysis of the Mn ore B&H record shows about 80 % share of unspecified amorphous phases, and the rest (crystalline phase) is mainly represented by quartz. Lepidocrocite, pyrolusite and hematite have a minor representation here as well.

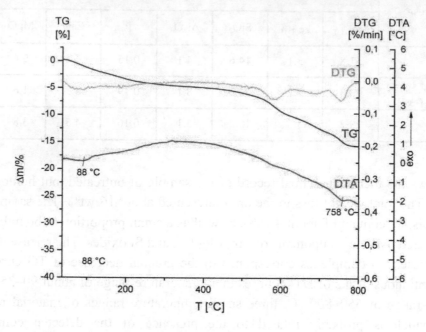

Fig. 86 Thermoanalytical record of Mn ore Turkey material sample

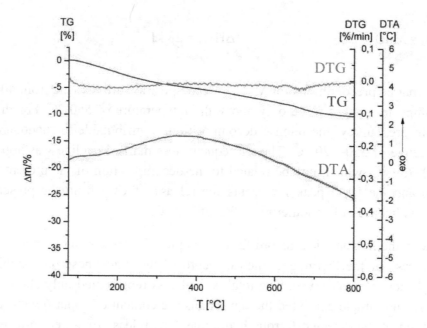

Fig. 87 Thermoanalytical record of Mn ore B&H material sample

Thermoanalytical record of homogenised but untreated Mn ore Bulgaria samples is shown in *Fig. 88*. The analysis of Mn ore shows relatively high water and carbon content. X-ray phase analysis shows a majority share of rhodochrosite and dolomite, which implies that it is a carbonate form of Mn ore. Thermogravimetric record shows a moisture removal (about 24 wt%) up to about 200 °C, indicating the presence of hydrate compounds. In the temperature range of 400–600 °C, about 10 % weight reduction of the sample was observed, which is associated apparently with the dissociation of $MnCO_3$ to MnO, which is then oxidised in the oxygen atmosphere to Mn_2O_3. Dissociation of dolomite present is likely to occur at temperatures higher than 800 °C. Given the temperature interval of sample measurement, this breakdown was not recorded.

$$4\,MnO + O_2 = 2\,Mn_2O_3 \qquad \Delta H_{298} = -89\ kJ \qquad\qquad (30)$$

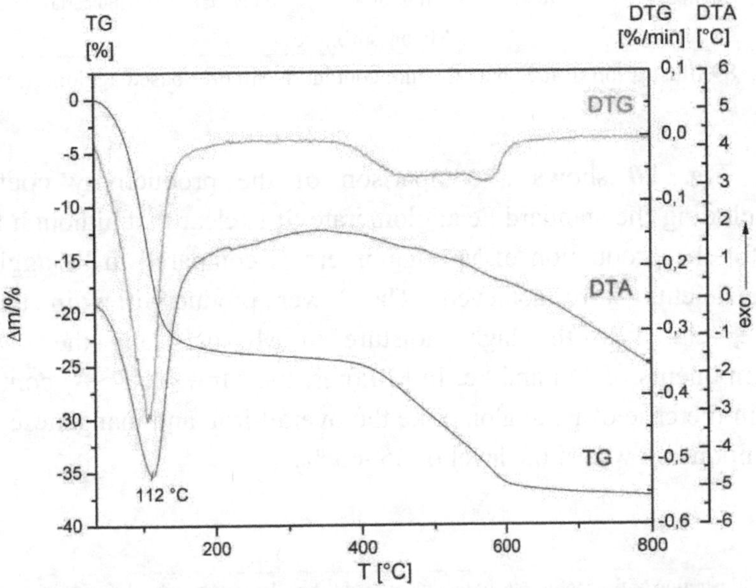

Fig. 88 Thermoanalytical record of Mn Ore Bulgaria material sample

In terms of economy of individual Mn agglomerate production, producing of Mn ore Bulgaria based agglomerates proves to be economically most difficult. When processing this Mn ore, the lower productivity due to worse rollability and high moisture content in agglomixtures has to be expected. Moreover, the economic costs of the agglomeration process will be also increased by a higher amount of fuel used to eliminate the high moisture and the higher carbonate content in Mn ore Bulgaria. With the increasing amounts of moisture in agglomixtures, the difference between the content of the coke powder of the dry and the actual state is linearly increasing, *Fig. 89*. In an agglocharge with high moisture content, slowing of vertical sintering speed due to the condensation of water vapour in the sintered layer may occur, resulting in the reduction of productivity. For the sintering of Mn ore with high moisture, it is necessary to allow for higher contents of fuel in the agglomeration charge [55, 56].

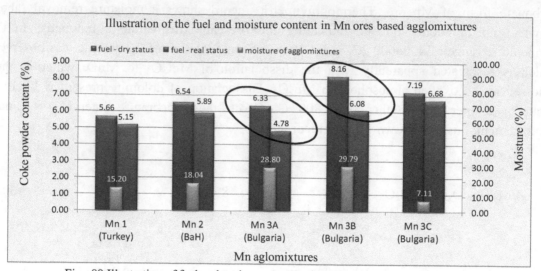

Fig. 89 Illustration of fuel and moisture content in Mn ores based agglomixtures

For illustration, *Fig. 90* shows a comparison of the productivity coefficient of Mn agglomerates, including the standard Fe agglomerate. It is clear that although higher levels of fuel were used for the production of Mn agglomerates compared to Fe agglomerate, lower productivity coefficients were achieved. The lower productivities in the case of Mn agglomerates are related to the high moisture in Mn ores and the lower content of metalliferous components of Mn and Fe. In Mn ores, the $Mn_{TOT} + Fe_{TOT}$ content was around 25–35 %, while in the case of Fe agglomerate the overall iron and manganese content of used metalliferous components was at the level of 55–65 %.

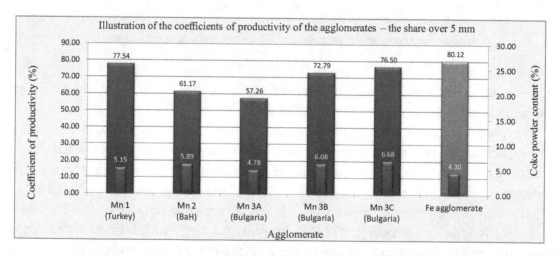

Fig. 90 Illustration of productivity coefficients of agglomerates – share over 5 mm

Fig. 91 shows the strength characteristics of the respective agglomerates. The highest strength index and conversely the lowest abrasion index was achieved for agglomerate Mn 1 (Turkey). The strength parameters of this agglomerate already approach those of the standard Fe agglomerate. The lowest strength index (+6.3 mm), and conversely the highest abrasion index (−0.5 mm), was measured for Mn agglomerate 3A (Bulgaria). In the case of the production of this agglomerate, there were very low temperatures in the sintered layer, related to the already mentioned high moisture in the input Mn ore Bulgaria, and the lower content of metalliferous components of Mn and Fe.

Normally, the temperatures of ferriferous material sintering (Fe ore or concentrates) are about 1200–1350 °C. In the case of poor Mn ores sintering, lower temperatures (about 900 to 1100 °C) were recorded. These lower temperatures are not automatically related to the lack of fuel in the agglomixture but may also relate to conditions of heat conduction and porosity of the sintered charge. In the case of lighter and more porous charge with a higher moisture content in the input agglopellets (typically poor Mn ore), lower temperatures may be recorded due to the water vapour condensation in the sintered layer and the higher rates of cooling of the agglomerate by sucked air. There are always micro-volumes in the sintered layer, where the temperature is higher than the measured temperatures on thermocouples.

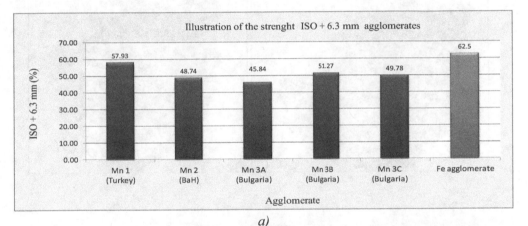

a)

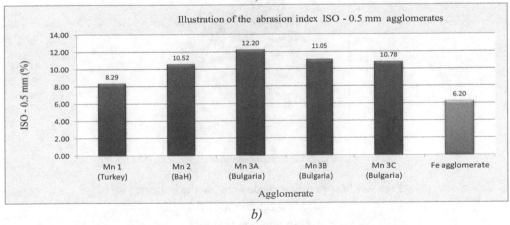

b)

Fig. 91 Illustration of strength characteristics of agglomerates
a) ISO +6.3 mm. b) ISO −0.5 mm

In all produced Mn agglomerates, manganese was mainly present in the lower oxides (at a low MnO content) or in complex compounds of manganese and iron lower oxides, which is positive for reducing properties of these input materials for FeSiMn production. **Fig. 92, 93** shows the final agglomerates of Mn, produced from poor Mn ores.

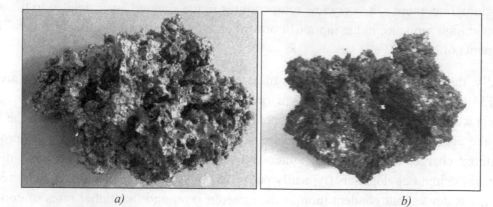

a)   b)

Fig. 92 Mn agglomerate made from Mn ore Turkey (a) and Bosnia and Herzegovina (b) [54]

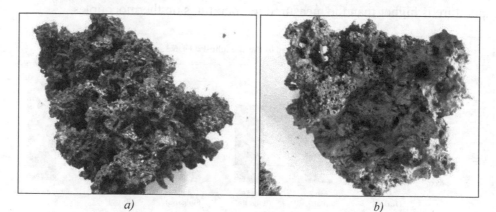

a)   b)

c)

Fig. 93 Mn agglomerate made from Mn ore Bulgaria [54]

a) Mn agglomerate 3A

b) Mn agglomerate 3B

c) Mn agglomerate 3C

7.3. Production of metallised agglomerate

Secondary raw materials produced in the whole cycle of metallurgical production are mostly valuable materials with a high content of iron and other usable components. On the other hand, in addition to valuable components, they contain a number of harmful elements and compounds that make it difficult to process them completely. One of the technological options is the production of the so-called metallised agglomerate. The metallised agglomerate production facilitates processing of, e.g., steelworks flue dust containing a high proportion of iron oxide, but also a high proportion of non-ferrous metals (especially Zn and Pb) [57–60]. The separation of non-ferrous metals from iron can be achieved by sintering such charges in the presence of a higher proportion of reducing agent.

In the metal production and processing, the equilibria between the condensed and gas phases are achieved in the furnace, while a variety of chemical reactions take place in the system, the course of which can be influenced to acquire different products. The changes in temperature, pressure or composition of the gaseous atmosphere can affect the balance between the condensed and gas phases, improving the parameters of some technological operations in metallurgical aggregates.

Reduction-oxidation processes take place in the presence of the gas phase with certain reduction or oxidative potential. The composition of the gas phase is crucial for the course of these reduction-oxidation processes. By the appropriate choice of both the total amount of fuel and the ratio of solid vs gaseous fuel in the agglomeration of steelworks flue dust, the CO/CO_2 ratio can be changed and conditions for the production of agglomerates with a higher content of Fe metal can be created, while containing less Zn and Pb.

For the production of the metallised agglomerate, coke powder is used as a fuel with similar characteristics as in the case of the production of other types of agglomerate (section 7.1.1). The amounts of used coke powder are significantly higher in the case of metallised agglomerate production than in the production of the standard agglomerate. The reason is the creation of a highly reductive atmosphere, in which it is possible to eliminate zinc, lead and other non-ferrous metals. Therefore, the term reducing agent is used in connection with coke powder in the production of metallised sinter.

A thermodynamic study and experiments were carried out for sintering of charges with an addition of flue dust from steelmaking process in the laboratory production of the metallised agglomerate. On the basis of thermodynamic analysis, a broad range of the amount of fuel used (6–27 %) was chosen, which should allow determining the optimum conditions for obtaining the material with required properties, appropriate chemical composition and low content of accompanying elements Zn and Pb. *Fig. 94* shows the change in the equilibrium composition of the system due to the increasing initial quantity of carbon in the system. With increasing amounts of carbon in the system (the amount of coke powder), the proportion of elemental Zn vapour increases as a result of the parallel decomposition of zinc ferrite and the reduction of ZnO. If the initial amount of carbon is increased above the limit of 6 moles, they become the majority zinc component of Zn $_{(g)}$ vapours.

Fig. 94 characterises the whole substance of non-ferrous metal separation within the agglomeration process, where the amount of carbonaceous fuel is an essential parameter affecting the efficiency of separation, especially Zn and Pb.

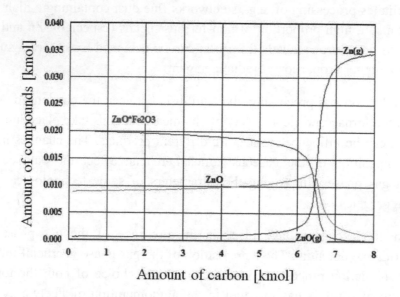

Fig. 94 Distribution of Zn as a function of carbon amount at 1300 °C
with maximum amount of air [60]

Fig. 95 shows that the degree of zinc removal increases with the increasing amount of fuel. As the amount of fuel increases, the $CO/(CO + CO_2)$ ratio increases as well, which characterises the highly reductive atmosphere, where the components containing Zn and Pb are reduced and escape in the gas phase of the flue gas. Not only coke, but also anthracite or different types of coal can be used as a reducing agent. The influence of the reducing agent type and its reactivity are ones of the most important factors influencing the whole process. It can be stated that the high reactivity of the reducing agent is not always convenient. The reduction of the fixed carbon begins to occur at temperatures of about 900–1000 °C and intensely proceeds only at temperatures above 1100 °C. If a reducing agent has higher reactivity, it may strongly react with oxygen in the gas phase even at temperatures below the temperature at which the reduction of iron oxides and oxides of non-ferrous metals by carbon occurs. In such case, a part of carbon is not used as the solid reducing agent. Very high temperatures (about 1300–1420 °C) were recorded in the case of metallised sinter production as a result of the already mentioned high content of coke powder (max. 27 %) in the agglomixture.

The production of metallised sinter can provide a product with a higher FeO (about 10–20 %) and Fe metal (about 5–10 %) content, and also low contents of Zn (max. 0.1 %) and Pb (max. 0.01 %). This product can be used not only as a charge for the BF but also for direct steel production. In such case, however, it is necessary to produce metallised agglomerate with a higher degree of metallization (min. 80 %).

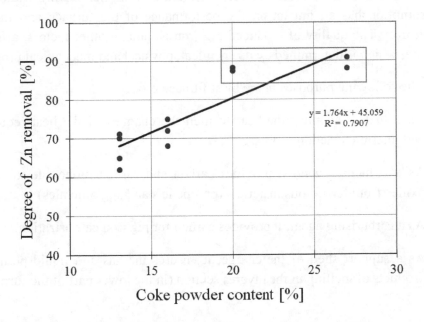

Fig. 95 Effect of fuel amount on the degree of Zn removal [59]

8. Use of conventional carbonaceous fuels in the pig iron production

Blast furnace coke is the crucial and the most important fuel additive in the blast furnace process. It is a refined fuel obtained by carbonisation of coal, i.e. heating without access of air. Its consumption has an impact on the performance of the furnace and the quality of produced pig iron. The quality of produced pig iron is, among other factors, affected by the quality of coke, which is determined by its chemical, physical and mechanical properties.

Coke serves the following purposes in the blast furnace process:

1. As a fuel, it supplies the heat to melt the charge and the heat required for the endothermic reaction processes.

2. As a reducing agent, it provides carbon and carbon monoxide to reduce metal oxides from ferriferous materials (ore, pellets and agglomerates).

3. As a carburising agent, it provides carbon for pig iron carburization.

4. As a support shell of the charge, it ensures the gas and liquid permeability for products of melting in the tuyeres section (in the lower part of the furnace).

Since coke is the most expensive component of the blast furnace charge and its production is associated with increased environmental impact, there is a continuous effort to reduce its specific consumption. Considering the roles of coke in the blast furnace process, it can be partially replaced with other fuels and reducing agents. However, it plays an irreplaceable role in ensuring the permeability of the charge. In recent years, a part of coke consumption has been reduced by the use of substitute fuels, namely:

- gaseous – heating gases, e.g. natural gas (calorific value of about $33-38$ $MJ.m^{-3}$), coke oven gas (calorific value of approximately $16-20$ $MJ.m^{-3}$),

- liquid – fuel oil (calorific value of about $38-40$ $MJ.kg^{-1}$), tar oil (calorific value of approximately $36-39$ $MJ.kg^{-1}$),

- solid – non-coking bituminous coal (calorific value of approximately $22-27$ $MJ.kg^{-1}$), *Fig. 96*, pea coke (calorific value of approximately $28-30$ $MJ.kg^{-1}$), coke powder (calorific value of about 28 $MJ.kg^{-1}$), adjusted plastics (calorific value of approximately $18-22$ $MJ.kg^{-1}$).

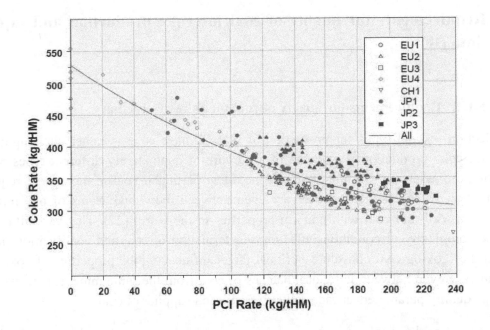

Fig. 96 Relationship between amount of blast furnace coke and injected pulverised coal [61] BF works (EU – the European Union, Ch – China, JP – Japan)

In addition to the traditional production of pig iron in blast furnaces, various alternative technologies of iron production have been developed around the world for decades. These technologies are implemented in different equipment, and various input materials are used for iron production. Alternative technologies of iron production are characterised by two main processes – smelting reduction and direct production of iron from ores.

Production of pig iron by smelting reduction is characterised by removing coke and the blast furnace as the basic production equipment from pig iron production. Bituminous coal is mostly used as a heat source and a reducing agent. Currently, COREX is the most widely operationally used technology. Other technologies such as HISMELT, DIOS, AISI, ROMELT and many others are either in the process of verification or are no longer used. However, the thermal efficiency of these technologies has not reached the efficiency of blast furnaces. Their advantages are mainly in investment and environmental areas, as processes of coke and agglomerate production are not used in these manufacturing technologies [62].

The direct iron production from ores is the technology of iron production in the solid state, i.e. at temperatures below the melting point. This technology uses other carbonaceous materials than coke for heating and reduction. Bituminous coal and natural gas are used for this purpose most often, from which reducing gas is produced. However, the share of these devices in the total production capacity is still small. Direct-reduced iron (DRI) or hot-briquetted iron (HBI) are used as a substitute for high-quality scrap steel in the steel production.

8.1. Requirements for quality of coal for coke production and injection into BF

8.1.1. Requirements for quality of coal for coke production

The selection of coal types for production of blast furnace coke is rather challenging. The charge for coke production is made up of a mixture of qualitatively different types of coal. Therefore, the assessment of possibility of including particular coal in the charge requires a number of tests. The coal properties vary even between individual seams of one mine and there is practically no coal with uniform properties in the world [27]. This variability stems from the complexity of conditions affecting the formation of coal and coalification process. Properties of coking coal constitute a critical factor influencing the properties of coke. Since they can vary even between individual batches of coal from the same mine, it is necessary to check the quality parameters at each delivery from each supplier of coal.

The quality of coal for coking is the most frequently simply assessed by the content of water, ash and volatile matter. The requirements for quality of coal for coke production differ among producers. **Tab. 20** gives the basic quality parameters of coal for coke production in Slovakia. The charge for the production of blast furnace coke generally consists of three types of coal, which are blended in certain proportions (e.g. 24 % of coking coal, 64 % of bituminous coal, and 12 % gas coal). The particle size of coal charge for the pouring operation is required in the range of 83 % of the grains below 3 mm (±2 % deviation).

Tab. 20 Basic qualitative parameters of coal for coke production

Type of coal	Water (W) [wt%]	Ash (A) [wt%]	Volatile matter (VM) [wt%]
Coke (highly coalified)	max. 12	max. 10	15–28
Medium (medium coalified)	max. 12	max. 10	24–35
Gas (low coalified)	max. 12	max. 10	30–36

The daily quality parameters of resulting coal charge are kept within the following ranges:

- water content (W) = 7.5–11.5 %,
- ash content (A) = 6–10 %,
- content of volatile matter (VM) = 24–28 %,
- swelling index (SI) = 6–8 %,
- vitrinite reflectance = 1.21 (65 % formed by the reactive components, and 35 % by inert components),
- dilatometry = min. 40 %, shrinkage = about 21 %,
- sulfur content = max. 0.40–0.80 %.

These qualitative parameters of coal charge are directly dependent on the quality parameters of individual coal types, and the operation affects them only by homogenization.

Water as a non-combustible component of coal reduces its utility value. Its content depends on the type of coal, the method of extraction, treatment, storage, and so on. Besides reduction of the calorific value, the water content in coal (W) affects its further processing, transport and storage.

Sulfur is present in the organic and inorganic matter of coal as pyrite, sulfate, elemental and organic. Although sulfur increases the high calorific value and low calorific value of fuel, it is always an undesirable component in coal used to make coke. During carbonisation of coal, most of the sulfur enters the coke, thus greatly reducing its value for metallurgical purposes.

Phosphorus comes from the original coal-forming plants, and its content in coal ranges from 0.01 to 0.11 %. During carbonisation of coal, phosphate salts do not decompose, so all the phosphorus passes into coke ash. Control of phosphorus content in coke is only possible by choosing coal from seams with the low content of phosphorus.

For coalification process characterisation and for classification of coal, it is necessary to understand its elementary analysis, i.e., the content of carbon, hydrogen, oxygen, sulfur, nitrogen and phosphorus and the heating value. The analysis of ash composition is also important. However, physical and rheological properties of coal are also significant, by which coking ability and other technological and technical characteristics of coal are assessed. Plasticity, swelling, expansion and shrinkage are the coking characteristics of coal used to assess the ability of coal to influence the course of carbonisation and the resulting quality of coke. All these features are described in detail in many professional publications focused on the characteristics of the charge for coke production (e.g. authors Kucková, Kret, Brož, Sabela, etc.) [7, 27, 38, 62] and are beyond the scope of this publication.

8.1.2. Requirements for quality of coal for injection into BF

The reduction of energy intensity of the pig iron production is a consequence of changes in the fuel base. This means using an increasing share of alternative fuels. Nowadays, coal is not used only for the production of blast furnace coke, but it is also used directly for blowing into a blast furnace in a powdered or granulated form by tuyeres. The new alternative processes for the production of pig iron (e.g. HISMELT) use coal for a similar purpose. The substitution of blast furnace (metallurgical) coke, which is the most expensive component of the charge, by blowing pulverised coal (up to 220 kg.t^{-1} p.i.) runs without a fundamental change of the blast furnace process technology. The theoretical maximum of this intensification technology is the value of 270 kg.t^{-1} p.i. This limit is imposed by carrying capacity of blast furnace coke and the thermochemical conditions in the blast furnace. In the case of pulverised coal injection, it is important to ensure improved coke strength after reaction with carbon dioxide (CSR) [63].

In blast furnaces, coal – usually of anthracite type – is used as a substitute fuel for expensive metallurgical coke. For the purpose of injection into the blast furnace, coal must have a specific particle size. Therefore, supplied coal of different granulometry up to 50 mm fractions must be ground. The grindability of coal is determined as Hardgrove Grindability Index (HGI). This property is very important especially for coal, which is used for injection into the blast furnace. When the HGI < 60, coal is difficult to grind. When the HGI > 60, coal has good grindability. The standard coal is coal from Somerset mines, USA. Its grindability is equal to 100. The calorific value of coal for injection into the BF should be approximately 28 kJ.kg^{-1}. Water content, ash content, sulfur and chlorine content are also important parameters of coal, which is injected into the BF.

The pulverised coal is transported to the furnace from a blowing vessel by collecting pipes through manifolds to individual nozzles, which are mounted in blowing fittings. The transport medium is nitrogen. The requirement for the quality of grinding is that 80 % of the pulverised coal should be ground to grain size below 0.09 mm. The moisture of the pulverised coal should be lower than 1.5 %.

Pulverised coal injection (PCI) into BF brings significant economic and technological advantages, e.g.:

- lower consumption of expensive coking coal and a wider range of used coal types,

- extended life of coke ovens,

- higher productivity and greater flexibility in the operation of the blast furnace,

- overall reduction of emissions from steel mills due to lower emissions from coke production,

8.2. Quality requirements for blast furnace coke

When reducing specific coke consumption, requirements for its quality, especially for resistance to mechanical and high-temperature stress, are increasing. Blast furnaces require coke with stabilised quality, minimum incombustible matter, suitable granulometry, high strength and low reactivity. However, the increase in coke quality leads to an increase in the cost of its production.

Coke quality requirements are derived from knowledge of the blast furnace coke degradation mechanism. Modern blast furnaces use coke (up to 100 kg/t of pig iron) in grain size "pea" (8 to 25 mm) along with the ore portion of the batch. This small coke has a larger surface area, readily reacts with carbon dioxide, and thus protects large pieces of coke from degradation. This allows injecting a larger amount of substitute fuels. Quality assessment of blast furnace coke is related to its chemical composition and physicochemical and mechanical properties.

The blast furnace coke grain size requirements are not uniform, typically 40 to 60 mm. High demands are placed on the uniformity of deliveries. It is unacceptable to charge coke with a grain size of less than 25 mm into the blast furnace, as it results in deterioration of the gas flow through the batch layer. The size of the largest pieces should not exceed twice the size of the smallest pieces.

The grain size composition of coke is evaluated according to the results of screen analyses, based on which the value of the medium grain of the coke is also calculated. According to this value, it is possible to predict a reduction of the coke grain size when passing through the blast furnace.

In practice, the assessment of chemical composition of coke involves the determination of water, ash, volatile combustibles, sulfur and phosphorus in coke. Recently, it has also been required to determine the chemical composition of ash, as the catalytic effect of ash constituents on the course and intensity of chemical reactions was clearly demonstrated [64]. The presence of alkali, sulfur and zinc reduces the initial oxidation reaction temperature. Changes in CaO, SiO_2, Al_2O_3 and Fe_2O_3 content lead to deterioration in the mechanical properties of coke as a result of increased coke debris and cracks. Coke consists of combustibles, ash, and water. Water and ash reduce the amount of combustibles, and hence, the utility value of coke as well.

The coke fuel contains on average 98 % of C, 0.1–0.5 % of H_2, 0.5–0.7 % of N_2, 0.4–1.1 % of O_2 and 0.5 % to 0.8 % of the volatile sulfur.

The content of volatile combustibles indicates the maturity of coke. The required content is from 0.5 to 1 %. Elevated values indicate deterioration of coke strength parameters and problems in the blast furnace off-gas cleaning system.

The water content of coke is related to the coke production and cooling technology. The minimum coke moisture should be at the level that prevents conveyor belt damage and excessive dust emissions. It is maintained within the range of 3.5 to 6 %, if possible.

Ash content and its composition have a major impact on coke consumption and furnace performance. Conversion of the acidic components of ash into a slag requires the addition of slag-forming additives where a one percent increase of ash in coke increases its consumption by 1.5–2.5 %, the consumption of limestone by 2 %, and thus the furnace performance drops. Although there are requirements for as low coke ash content as possible, the content up to 12 % is considered satisfactory.

Alkali, sulfur and phosphorus content is detrimental to the furnace operation and pig iron composition [62, 64]. The required sulfur content in coke is 0.5 to 0.9 %. The required phosphorus content is 0.02 to 0.06 %, and the required alkali content in ash is 2 to 4 %. The content and composition of ash in coke is determined by the quality of batch coal used in its production.

The composition of coke is related to its calorific value, which is calculated per combustible, on average about 30–33 $MJ.kg^{-1}$, for blast furnace coke.

The rate of grain degradation through a blast furnace passing depends on physical-chemical and mechanical properties of coke under thermal and physical-chemical process conditions. It is related to the structure of coke, i.e. the size, shape and distribution of pores in the coke, coke texture, and ash content and composition. These factors influence the strength of coke, its reactivity, conductivity, crackiness, lumpiness, resistance to breakage at rapid temperature rise, and so on [27].

The mechanical strength of coke is its resistance to abrasion and impact. To assess the coke strength parameters, tests based on subjecting certain amount and size of coke grains to the mechanical stress in drums or dropping coke from a certain height on a steel plate are commonly used in our country and abroad. Parameters of strength and abrasion determined by the coke drum test MICUM have become the main indicator of coke quality in our country.

The strength of coke M40 expresses the grain coke fraction above 40 mm after the test in the drum, and the gross coke weighed into the drum before the test, expressed as a percentage. This value should be as high as possible. High-quality coke achieves the strength of approximately 80–85 %. The value for quality coke 70–80 %.

The abrasion of coke M10 is a fraction of coke grain below 10 mm from the total coke of the drum test, expressed as a percentage. This value should be as low as possible. The appropriate coke abrasion is in the range from 6 to 8 %.

For assessing the coke grain degradation at the bottom of the blast furnace, the NSC test was adopted as well as the parameter of coke strength after the coke oxidation reaction with CO_2 – CSR and the coke reactivity CRI. These assessments have been added to the basic indicators of coke quality for modern blast furnaces. It is especially true for BFs that use substitute powder fuels. The CSR coke strength is determined by the drum test as a fraction of coke grain size above 10 mm after the drum test, and the weight of coke weighed into the drum, expressed as a percentage. CSR strength typically ranges from 52 to 62 % and should be as high as possible. CRI coke reactivity is determined from coke weight loss before and after the test. It usually varies from 27 % to 37 % and should be as low as possible. It is increased by the catalytic effect of ash constituents. In *Fig. 97*, a blast furnace coke microstructure is shown [65]. Coarse anisotropic components (optically active components with a larger surface) undergo the oxidation reaction and heat shock to a smaller extent than fine anisotropic and isotropic components. Isotropic texture is mainly typical for highly reactive coke samples.

The quality of produced coke can be approximately estimated already by its appearance. The silver-grey colour is a sign of well-matured coke. The black coke is insufficiently mature, soft and easily abraded. The strength of coke can be judged by the degree of cracking and the sound of its fall on a hard plate, which is ringing for hard coke. It is possible to assess the size of pieces and the uniformity of sorting. The appearance of the macrostructure gives an overview of the porosity of coke, and the light grains in coke show the amount and distribution of ash.

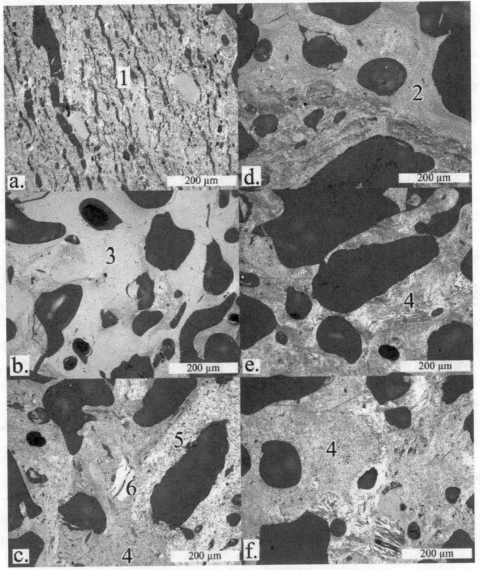

Fig. 97 Blast furnace coke microstructure [65]

(a–d) highly reactive coke, (e–f) low reactive coke
1 – isotropic texture (non-fused grains), 2 – isotropic texture (fine mosaic),
3 – isotropic texture (melted grains), 4 – anisotropic texture (fine mosaic),
5 – anisotropic texture (coarse mosaic), 6 – anisotropic texture (non-fused grains)

For blast furnaces in Slovakia, the following quality requirements for blast furnace coke were established based on the knowledge of raw material (coal and ores) import possibilities and the state of technical equipment in the Coke Plant and Blast Furnaces divisions:

- water content = max. 4–6 %,

- ash content = max. 10.1 %,

- volatile combustible matter = max. 0.62 %,

- sulfur content = max. 0.58 %,

- medium grain size = max. 57 mm,

- coke strength M40 = min. 78 %,

- coke abrasion M10 = max. 6 %,

- CRI reactivity = 28–32 %,

- coke strength CSR = 54–57 %.

In *Tab. 21*, there is a comparison of average blast furnace variables in Japan and the European Union. The total amount of carbonaceous fuels used in BF plants in the EU is within a narrower interval than in the case of Japan. With a comparable amount of injected powdered coal, the amount of blast furnace coke used is lower in the case of BF plants in the EU. It is also in relation to the coefficient of production which is higher in the EU. These parameters indicate that the blast furnace process is more efficient in the EU – also because of better input materials used in the EU in recent years. The number of tailings and their composition in ferriferous materials have a major impact on coke consumption and furnace performance.

Modern blast furnace technologies employ the injection of a large number of substitute fuels (along with oxygen) into the hearth of a blast furnace, *Fig. 96* [61]. This technology will make it possible to reduce the coke fraction in the blast furnace by half in the near future. Such a reduction in the proportion of coke in the upper batch layer causes a significant deterioration of the gas dynamics conditions in this blast furnace section. A solution to this situation is to move a part of the ore charge from the shaft to the bottom of the blast furnace by blowing the iron oxide dust through a tuyere. The situation can also be improved by loading the part of coke with the grain size of "pea" to "nut" along with the ore part of the charge. This part of coke should be more reactive.

Tab. 21 Comparison of blast furnace indicators

Indicator	SI unit	Japan	EU
Temperature of blown wind	[°C]	1157	1178
Volume of enriched oxygen in blown wind	[Nm³/t p.i.]	40	34
Volume of blown wind	[Nm³/t p.i.]	980–1190	990–1150
Total amount of fuel	[kg/t p.i.]	510–570	470–530
Amount of blast furnace coke	[kg/t p.i.]	360–420	320–380
Amount of injected pulverised coal	[kg/t p.i.]	150	150
Adiabatic temperature of flame	[°C]	1920–2230	1980–2240
Amount of slag	[kg/t p.i.]	270–300	270–300
Production coefficient	[%]	74.4	82.7

9. Use of traditional carbonaceous fuels in the production of steel and ferroalloys

Carbonaceous fuels (e.g. anthracite, coke and graphite) are mainly used as carburisers in steelmaking process facilities (oxygen converter and electric arc furnace). To increase the yield of steel produced in the electric arc furnace, mainly anthracite is used, which also acts as a reducing agent (e.g. in the reduction of FeO from steelmaking slag). To increase the efficiency of an electric arc furnace, fuel burners use mainly hydrocarbon fuels (e.g. natural gas), but also anthracite and coal. Carbonaceous materials are also used to produce electrodes for electric arc furnaces.

The production of ferro-alloys is based on reduction processes, just like the production of pig iron in the blast furnace. The efficiency and quality of ferroalloy production in electric arc furnaces is very much dependent on the quality of the reducing agents used for this production. Practically, bituminous coal and coke are used as major carbonaceous reducers for the production of ferro-alloys. Similarly to the production of steel in the electric arc furnace, carbonaceous materials are used to make electrodes for the production of ferro-alloys.

9.1. Requirements for the types and quality of carbonaceous materials for steel production

Carbonised materials have multiple uses in the production of steel in main steel aggregates (oxygen converter and electric arc furnace). The main use of these carbonaceous materials is coarse carburization during steel melting (in all steel aggregates), frothing of the slag during steel melting in the electric arc furnace, fine carburization in the ladle treatment of steel, and for casting powders for continuous casting of steel. In steel production, carbonaceous materials modify the heat balance of the steel process utilising thermal energy from carbon exothermic reactions with oxygen. Carbonaceous materials are also an important part of refractory materials used to line steelmaking aggregates. For example, magnesite-carbon material or fire clay-graphite refractories. In the electric arc furnace, graphite electrodes are used, which are made from graphitic carbon. The basic physicochemical properties of the carbonaceous materials used as carburisers in the production of steel are given in *Tab. 22*. In industrial practice, calcined anthracite is used the most, *Fig. 98a*.

Tab. 22 Physicochemical properties of carbonaceous carburisers for steel production

Parameter	Unit	Calcined anthracite	Calcined petrol coke	Calcined pitch coke	Graphite carburiser	Coke
Content of carbon	[%]	min. 95	min. 98	min. 94	min. 98	min. 85
Ash content	[%]	max. 4	max. 0.8	max. 5	max. 0.8	max. 10
Content of volatile matter	[%]	max. 0.8	max. 0.5	max. 0.7	max. 0.8	max. 0.8
Moisture	[%]	max. 1	max. 0.5	max. 0.5	max. 0.5	max. 4
Content of sulfur	[%]	max. 0.3	max. 0.8	max. 0.3	max. 0.1	max. 0.5

In addition to standard carburisers, materials made from undersize fractions of high-carbon materials are used. Typical representatives are, for example, graphite pellets made from pure graphite sub-fractions, *Fig. 98b*. Graphite pellets have a carbon content of min. 88 %, ash content of max. 0.8 %, volatile flammable content of max. 1 %, and sulfur content of max. 0.1 %.

a)   b)

Fig. 98 Carbonaceous carburisers for steel production
a) calcined anthracite, b) graphite pellets

Graphite electrodes are designed for electric arc furnaces for steel production but also for pan furnaces used for the refining of crude steel in batch ladle processing technology. Special requirements are placed on them, especially with respect to chemical purity, strength and high conductivity (specific electrical resistance as low as possible), *Tab. 23*. Individual designations of graphite electrodes are related to the quality class, e.g. UHP electrodes are the highest quality electrodes that are used in the most powerful electric arc furnaces with the power output over 100 MW.

Tab. 23 Characteristics of graphite electrodes for steel production

Parameter	SI unit	Graphite electrode RP	Graphite electrode HP	Graphite electrode SP	Graphite electrode UHP
Ash content	[%]	0.30	0.20	0.15	0.10
Density	[g/cm^3]	1.56–1.60	1.62–1.66	1.65–1.69	1.69–1.75
Porosity	[%]	25–28	23–25	22–24	20–23
Electric resistance	[μΩm]	8.0–9.5	5.6–7.0	5.5–6.5	4.5–5.9
Strength	[MPa]	8.0–10.0	10.0–12.0	11.5–13.0	12.5–15.0

Natural and synthetic materials containing at least 90 % carbon are the starting materials for the production of graphite electrodes. They include anthracite, graphite, bituminous coal coke, pitch coke, petroleum coke, return materials from the production of electrodes, while bituminous coal electrode pitch or tar is used as a binder. Carbon graphitisation itself (change of carbon modification) takes place at temperatures of about 2600 to 2800 °C, which ensures high electrical and thermal conductivity of an electrode, increases the heat resistance, improves the mechanical properties, and reduces the tendency of electrodes to oxidation. *Fig. 99* shows graphite electrodes and their location in the electric arc furnace during steel melting.

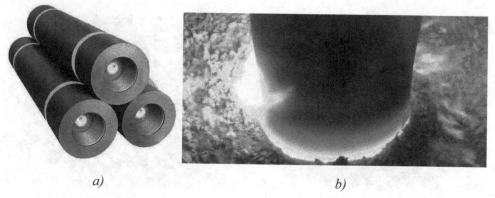

a)   *b)*

Fig. 99 Graphite electrodes for steel production (a) and electrode location in furnace (b)

More than 95 % of the world electric steel production is produced using graphite electrodes. Only a very small part of steel is produced in electric arc furnaces using carbon electrodes without graphitisation. The average electrical resistance of the graphite electrodes is 4 to 5-times higher, and the current density is up to 2.5-times higher, so a high concentration of power can be achieved, which is advantageous in terms of good efficiency and high temperature. Graphite electrodes have a lower weight, so they are easier to handle and break less often in comparison with carbon electrodes, which reduces the electrodes consumption in steel production. The consumption of electrodes depends on the type of electrodes, the melting regime, furnace size and the method of sealing electrode openings at the cover of the furnace, as well as the perfect closure of working holes. The electrodes are worn by oxidation by air oxygen entering the furnace, dissolving in slag during the boiling period of the bath, electric arc sputtering, and by mechanical manipulation.

9.2. Requirements for types and quality of carbonaceous materials for the production of ferroalloys

The majority of ferroalloys are currently produced electrothermally – either in a reducing or refinery electric arc furnace. Reducing electric arc furnaces are used for carbothermic ferroalloy production, i.e., carbonaceous reducing materials are required for reduction processes. The efficiency and quality of ferroalloy production in electric arc furnaces is very much dependent on the quality of reducing agents used for the production. In practice, the hard medium-coalified coal, coke and semi-coke are the primarily used carbonaceous reduction agents for the production of ferroalloys at present. From the standpoint of ferroalloy production, it is necessary to use a type of reducer (whether on the basis of coal or coke) with the lowest content of ash, sulfur and phosphorus. At the same time, these reducers should have high reactivity, high electrical resistance, sufficient strength, optimum size, good gas permeability and high thermal stability.

The most important properties of coal and coke used in the production of ferro-alloys include reactivity. In general, the reactivity of coal and coke decreases with the increasing degree of coal enrichment. The coke reactivity decreases with the increasing percentage of carbon. The smaller coke is, the more reactive it is. Smaller and more porous coke is more reactive than dense coke. Coke reactivity also depends on the composition of coal batch and the coking conditions. Increasing the coking temperature and longer carbonisation times reduces the coke reactivity except for certain cases. Coke reactivity decreases with the decreasing porosity and the increasing ash content. The coke reactivity increases with the increasing ash content, if the ash contains a larger amount of iron oxides and alkali metal compounds.

Inferior quality types of coal – made from a higher gas coal fraction batch with higher porosity and higher volatile combustible content – are the most appropriate for the ferroalloy production out of the conventional metallurgical coke types used for the production of pig iron. The higher content of volatile matter in the reducer increases the amount and combustion heat of flue gases in the production of ferroalloys. The so-called reactive coke containing approximately 2–4 % of volatile matter is suitable.

Bituminous coal with varying degrees of coalification generally has a higher reactivity and higher electrical resistance than metallurgical coke. Charcoal, semi-coke and petroleum coke (crude oil coke) also have better physicochemical properties than metallurgical coke.

The higher electrical resistance of the reducer is also needed to achieve the higher electrical resistance of the whole batch and more efficient use of heat in a reducing electric arc furnace. The electrical resistance of the reducers decreases with an increase in the carbon content of the reducers, and on the contrary, increases with the increase of the volatile fuel and ash content in the reducer.

Several types of carbonaceous reducers are currently used for the production of ferroalloys in the EAF. For each type of ferroalloy produced, different types and ratios of reducing agents may be used. However, in principle, the most widely used carbonaceous reducing agents are pea coke, nut coke, bituminous coal and anthracite. Many plants also use wood chips, whose main task is to increase the permeability of the charge. In addition to traditional reducers, other types are also used for reduction, e.g. higher quality brown coal (with the ash content as low as possible), semi-coke, charcoal and petroleum coke, **Tab. 24.**

For the production of high-quality ferroalloys, it is necessary to use reductants with the lowest content of ash (mainly Al_2O_3 components), sulfur (its content does not have to be a critical factor), and phosphorus (it is a critical factor, as it is not possible to achieve a higher degree of ferroalloy dephosphorization in reducing electric arc furnaces).

Tab. 24 Typical analysis of reductants used in production of ferroalloys

Material	Ash [%]	Volatile matter [%]	C [%]	S [%]	P [%]	Analysis of ash				
						SiO_2 [%]	Al_2O_3 [%]	FeO [%]	CaO [%]	MgO [%]
Brown coal	5–15	30–50	30–40	0.8–1	0.080	62.80	19.80	4.96	1.82	5.54
Bituminous coal	3–8	28–35	65.00	0.85	0.030	45.56	28.50	8.54	2.57	6.45
Anthracite	7.20	5.70	87.10	0.70	0.009	45.20	29.20	10.35	2.52	5.34
Coke	12.90	1.34	85.32	0.56	0.020	56.92	22.30	6.68	1.82	6.65
Semi-coke	7.20	5.54	83.23	0.45	0.007	45.65	23.21	11.45	12.43	3.43
Charcoal	1–5	10–20	70–80	0.30	0.020	43.60	10.33	3.99	32.00	5.00

In Norway, types of bituminous coal with varying degrees of coalification were tested in ferroalloy production (mainly ferrosilicon) [31]. The best results have been achieved with the use of the U.S. coal Blue Gem, which has very low ash and phosphorus content. It is low-coalified coal (R_O = 0.7–0.8) with a high proportion of reactive vitrinite and the minimum of anisotropic parts. The Blue Gem coal had a high reactivity (in the so-called SiO test, where the amount of unreacted SiO in mL was evaluated; the lower value in the test means that more $SiO_{(g)}$ was reduced). The Australian Peak Downs coal also had good reactivity, although it was coal with a higher degree of coalification (R_O = 1.3–1.5). Nevertheless, it had a high content of vitrinite and fine isotropic parts, **Tab. 25.**

Tab. 25 Characteristics of bituminous coal [31]

Coal	Moisture [%]	Ash [%]	Volatile matter [%]	S [%]	P [%]	Al_2O_3 [%]	Reactivity of SiO test [mL SiO]	Reflectance of vitrinite R_o [%]
Blue Gem (USA)	2.4	1.5	36.0	0.60	0.003	0.33	765	0.70–0.80
Peak Downs (Australia)	0.9	11.6	19.8	0.58	0.057	3.30	749–797	1.30–1.50
Staszic (Poland)	2.5	5.9	32.0	0.47	0.029	1.44	854	0.65–0.70

In order to achieve an efficient technological process for the production of individual types of ferroalloys, it is necessary to comply with the requirements for the reduction agents, which are summarised in the following text.

A. Requirements for coke:

• For the supplied coke, it is necessary to comply with the specified amount of oversize and undersize fractions at the given limit of grain size. As to the qualitative parameters of coke, the content of ash, water, volatile combustible, sulfur and phosphorus, its combustion heat, and calorific value are given according to agreement (according to the application). The strength parameters M25 and M10 (or I20 and I10), reactivity and the electrical resistance is normally not determined for lower grain classes (pea and nut).

• Since coke is produced in most enterprises for metallurgical purposes, its production is governed by requirements for the quality of coarse grain sizes, out of which about 65 to 75 % is separated. In the most ferroalloy production technologies for electric arc furnaces, pea coke with the grain size of 10–25 mm (or nut coke with the grain size of 20–40 mm) is used. With the decreasing grain size in coke, water and ash content increases.

• The required water content in coke is low. It should not be higher than 8 % for dry coke.

• The ash content is required as low as possible because it has an impact on the amount of slag, thermal balance of electric arc furnaces, furnace performance and consumption of coke. High-quality coke has an ash content of about 8–10 %, while the content up to 12 % is considered acceptable.

• The sulfur content is required as low as possible, as sulfur passes into metal, which it degrades. Quality coke has the sulfur content of 0.6 to 0.9 % (the stricter requirement is max. 0.5 %). Also, the phosphorus content is required as low as possible. Quality coke has the phosphorus content of about 0.01 to 0.1 % (the stricter requirement is up to 0.02 %).

• The cold coke strength (expressed by the Micum and Irsid tests values) M25 and I20 should be above 70 %, and the maximum abrasion of coke M10 or I10 should be below 10 %.

• The reactivity of coke for ferroalloy production, expressed as CRI, should be above 30 %, preferably about 35 %. Reduction agents with reactivity as high as possible (higher than 1.0 mL/g.s or 1.0 cm^3/g.s are required for ferroalloy production – abroad, reactivity is sometimes determined by the CO_2 consumption per unit of time).

• The specific electrical resistance is one of the important properties that influence the production of ferroalloys. The productivity, specific power consumption, working voltage, electrode storage and other production indicators depend on electrical resistance of coke. The electrical resistance should be as high as possible – greater than 0.01 Ω.m.)

B. Requirements for bituminous coal:

• Grain size and water content properties should be close to the values required for coke properties. The ash content should be up to 10 % (the stricter requirement is max. 8 %).

• The content of volatile matter should be about 20–30 % (30–35 % in less coalified coal). For ferro-silicon production, less coalified coal with higher reactivity and higher specific electrical resistance is preferred.

• Phosphorus and sulfur content should be as low as possible.

• The reactivity and electrical resistance should be as high as possible – generally above the level for coke (above 35 % or above 0.01 Ω.m.).

Similarly to the production of steel in an electric arc furnace, carbonaceous materials are used to produce electrodes for the production of ferroalloys. Self-sintering electrodes (for carbothermic production) and graphite electrodes (for metallothermic production and ferroalloy refining) are used the most. The graphite electrodes are manufactured using a special technology of low-ash graphite and tar. They surpass other types of electrodes in all utility properties. The graphite electrodes should have the lowest possible electrical resistance, high current density, high strength, low ash content, and high graphite carbon content. These electrodes are the most expensive, which hinders their greater use in the large volume production of ferroalloys. They are used only in cases where self-baking electrodes cannot be used because of the current load and the product quality, especially due to the carbon content. Allowed current density of the graphite electrodes is several times higher than with the self-baking electrodes (the value for the graphite electrode is in the range of 20–30 A.cm^{-2} according to the diameter and the quality of the electrode; the value for the self-baking electrode is approximately 6–7 A.cm^{-2}).

For the production of both carbon ferroalloys (e.g. FeMnC and FeCrC) and silicon ferroalloys (e.g. FeSi, FeSiCa and SiCa), the above self-baking electrodes are used. The self-baking electrodes are mostly of Söderberger type and consist of steel cylinders filled with electrode paste. Electrode paste is a dry mixture of calcined anthracite, pitch coke or petroleum coke with the specific granulometry (75–80 %) and pitch tar binders with the specific softening point (20–25 %). Special requirements are placed on them, especially with respect to chemical purity, strength and high conductivity (or as low as possible specific electrical resistance), *Tab. 26*, *Fig. 100*.

Tab. 26 Typical properties of electrode paste for production of ferroalloys [67]

Parameter	SI unit	EH 1	EH 2	EH 3	EH 4	EH 5	EH 6
Ash content	[%]	6	4	3.5	4	5	4
Volatile combustible matter	[%]	12.5	12.5	14.5	12.0	10.0	11.0
Porosity	[%]	24.5	26.0	25.5	26.0	22.0	24.0
Plasticity	[mm]	53–68	55–70	55–65	55–70	60–75	58–72
Electric resistance	[$\mu\Omega$m]	100	110	65	85	65	75
Compressive strength	[MPa]	20	20	17	18	18	20

Fig. 100 Carbon electrode paste for production of ferroalloys

10. Traditional carbon fuels and the environment

Industrial production of iron, steel and ferroalloys is material and energy intensive, and at the same time, it has a negative impact on the environment. More than half of the material inputs are on the output released in the form of waste gases and solid waste (secondary products) [68]. Most of the emissions that enter into the atmosphere come from carbonaceous fuels. At present, in many parts of the world (e.g. China and South America), we are witnessing such pollution that does not only endanger the environment but also directly threatens the health and lives of people. The process of ecologisation of metallurgical technologies is currently actively pursued only in Europe, North America, Southwest Asia, and in Japan (e.g. ULCOS projects – reducing CO_2 emissions in iron and steel industry by 50 % by 2015). Systematic pressure on manufacturers has brought a substantial improvement through the gradual modernisation of metallurgical technologies and equipment. Minimization of emissions and solid waste needs to be addressed globally (i.e. also in regions where emissions are not yet an obstacle as they are not subject to stringent limits).

Gaseous, liquid and solid substances emitted into the air are called emissions. Emissions that have a negative impact on the environment (including human health) are called pollutants. Pollutants can be introduced directly into the atmosphere, but they may also be formed as a part of the physical or chemical transformation. Pollutants that are directly introduced into the air are referred to as primary emissions. Pollutants that are generated in the atmosphere in the course of physical and chemical reactions due to the effects of the Sun and water are referred to as secondary emissions.

The maximum allowable quantities of pollutants discharged from the source of pollution into the atmosphere indicate the so-called emission limit, expressed as a mass concentration of a pollutant in the exhaust gas ($mg.m^{-3}$) or volumetric concentration in ppm of gaseous pollutants.

Pollutants that are dispersed in the ground-level atmosphere (i.e. in the air we breathe) are called imissions. The maximum permitted concentrations of these substances are also limited by the legal standard in the form of imission limits. Air quality standards are given in weight concentrations, i.e. in units of mass per unit volume of air at the temperature of 273.15 K and the pressure of 101 325 kPa. The longer the exposure time of certain concentration of a pollutant in the ambient air, the lower the permitted concentration of the substance in the atmosphere. The maximum short-term concentration of a pollutant in the ambient air is expressed as a one-hour maximum concentration ($mg.h^{-1}$). For practical use and regulation of air pollution sources, the maximum allowable 24-hour concentrations ($mg.day^{-1}$) are used. Pollutants from the atmosphere that fall to the Earth's surface are called depositions. Depositions are expressed by two components, namely dry and wet deposition. The dry deposition is the amount of the fallout in an untransformed form (fly ash). The wet deposition consists of pollutants resulting from the conversion in the atmosphere.

The essential, i.e. the most closely monitored contaminants, include particulates, carbon dioxide, carbon monoxide, nitrogen oxides, carbon dioxide, organic gases and vapours as total organic carbon, polychlorinated dibenzo-p-dioxins (PCDDs) and polychlorinated dibenzofurans (PCDFs) [69]. All of these pollutants also come from the use of carbon fuels (e.g. combustion processes, oxidation processes, reduction process, carburising, etc.). Other pollutants such as heavy metals, fluoride, halogen compounds, unburnt hydrocarbons and non-methane volatile organic compounds are then emitted into the atmosphere by fuel combustion in lower amounts. Because of the toxicity and persistence, these pollutants have a very negative impact on the vegetation, animals and people.

Particulates (PM)

The solid (dust) particles in the atmosphere constitute a mixture of compounds consisting of carbon, dust and aerosols. These substances are also formed in the context of metallurgical production of iron, steel and ferroalloys. Their main source is the combustion of solid fuels (especially coal and coke) or the entrainment of dust from the furnace apparatus to a flue gas exhaust system. A small proportion of dust may contain very small particles formed by the condensation of compounds that gradually evaporate throughout the metallurgical technology. This creates, for example, soot, which can be very corrosive. The combustion of liquid fuels is the source of particulate emissions well, yet to a much lesser extent than, for example, coal and coke. In many devices, fugitive emissions can also occur that are released into the atmosphere primarily by vents, leaks in piping and appliances, handling and storage of coal in open space, and handling of ash.

Carbon dioxide (CO_2)

Carbon dioxide is the main reaction product of the combustion of all fossil fuels, and its emissions are directly proportional to the carbon content of fuels. The carbon content in brown coal, lignite, bituminous coal and coke varies from about 60 to 87 % of weight. The carbon content in gas oil and fuel oil is about 85 %. Within the metallurgy of iron, steel and ferro-alloys, coal with medium and higher degree of coalification and coke (carbon content is about 80–87 %) are used the most. However, the lower carbon dioxide emissions are produced from gaseous fuel (e.g. natural gas with the carbon content of about 70–75 %). In recent years, carbon dioxide has been increasingly attributed the largest contribution to the changes of climatic conditions on Earth. On the other hand, carbon dioxide is a normal part of Earth's atmosphere and is essential for the existence of life on Earth. However, if the amount of CO_2 in the atmosphere is above the limit, there is a lower amount of vitally important oxygen in the air. It disrupts the balance in nature and the ability of plants to process these over-limit amounts of carbon dioxide by photosynthesis.

Carbon monoxide (CO)

Carbon monoxide is the main product of incomplete combustion of fossil fuels. However, in the context of metallurgical processes, it may also be a product of oxidation and reduction reactions that take place in the production of iron, steel and ferroalloys. Carbon monoxide in the combustion process is an indicator of the risk of corrosion and incompletely burned fuel, which obviously has a significant impact on efficiency. Carbon monoxide is also a part of the increasingly recovered secondary metallurgical gases, which are used as fuel, for example, coke oven gas, blast furnace gas, etc. Currently, there are also produced synthesis, pyrolysis and water gases, which are obtained by pyrolysis or gasification of solid fossil fuels. Since the synthesis and pyrolysis gases are further used (including metallurgy), their environmental impact is minimal.

Nitrogen oxides (NOx)

The principal nitrogen oxides produced during the combustion of fuel and consequently emitted into the atmosphere are the nitrogen monoxide (NO), nitrogen dioxide (NO_2) and nitrous oxide (N_2O). The first two oxides are known as a mixture of nitrogen oxides (NO_X). More than 90% of nitrogen oxides, which originate in the context of metallurgical technology, is emitted in the form of monoxide. However, the gas is rapidly converted into nitrogen dioxide in the atmosphere, and it further changes into the nitric acid, which is combined with moisture, leading to the formation of acid rain. Thermic oxides of nitrogen originate from the reaction between oxygen and nitrogen from air. The fuel nitrogen oxides are formed from the nitrogen contained in the carbonaceous fuel, and instantaneous (fast) oxides of nitrogen are formed by conversion of molecular nitrogen in the front part of the burner flame in the presence of intermediate hydrocarbon compounds [69]. The formation of thermic nitrogen oxides is highly dependent on temperature. If the combustion occurs at the temperature below 1000 °C, the nitrogen oxide emissions are considerably lower. At the flame temperature below 1000 °C, the formation of NO_X depends mostly on the content of nitrogen in the fuel. Formation of thermal NO_X is dominant especially in devices that use gas and liquid fuels.

Sulfur oxides (SO_X)

Sulfur oxides (SO_X) are found in various forms and produce several pollutants. The most significant harmful emission is sulfur dioxide (SO_2). Acid aerosols may be formed by oxidation of sulfur dioxide in the atmosphere. Sulfur oxides come mainly from sulfur present in the fuel. Fossil fuels contain sulfur as inorganic sulfides or organic compounds. In coal, sulfur is present as pyritic sulfur, organic sulfur, sulfur in the form of salts, and as elemental sulfur [69]. During combustion, the largest amount of sulfur is released as sulfur dioxide (SO_2). Sulfur dioxide (SO_2) resulting from the combustion of traditional carbon fuels containing sulfur is included in the typical and most common components of emissions that pollute the air in the context of metallurgical technologies, together with solid particles. The major issue related to the burning of coal and coke is associated with sulfur. Certain types of lower quality coal contain 2–5 % of sulfur, which is oxidised during combustion to sulfur dioxide (SO_2). Sulfur dioxide present in the wet air is very soluble in water, and then sulfuric acid is formed, which is a part of the so-called acid rain.

Total organic carbon (TOC)

Total organic carbon (TOC) is the total amount of carbon in organic substances, water and gases. TOC may also originate from the thermal treatment of carbonaceous fuels. Total organic carbon includes a wide variety of hydrocarbons, and these include the persistent organic pollutants (POPs), including, e.g., non-methane volatile organic compounds (VOCs), most of the polycyclic aromatic hydrocarbons (PAHs), polychlorinated biphenyls (PCBs), polychlorinated dibenzo-p-dioxins (PCDDs) and polychlorinated dibenzofurans (PCDFs) [69].

Volatile organic compounds (VOC)

Volatile organic compounds (other than methane) can be defined as compounds of carbon except for CO, CO_2, H_2CO_3, metal carbide, metal carbonates and ammonium carbonate. Generally, the volatile organic compound (VOC) emissions represent a large group of easily evaporating liquids, which comprise a variety of organic chemicals. These substances enter the environment primarily from combustion processes during the incomplete combustion of fossil fuels. The most important compound with a negative impact on the environment from the class of organic compounds are xylenes, toluene (C_7H_8) and benzene (C_6H_6) [69].

Polycyclic aromatic hydrocarbons (PAHs)

Fossil fuels (mainly coal and natural gas) are composed of hydrocarbons which are converted into gaseous components (carbon dioxide and water vapour) during combustion. Since the combustion process is always accompanied in part by incomplete combustion, other components such as carbon monoxide and carbon in solid form (which is a part of smoke) are originate in the chemical reactions as well. Polycyclic aromatic hydrocarbons are also produced by incomplete combustion of coal. The polycyclic aromatic hydrocarbons (PAHs) represent a very large variety of different substances, characterised by the content of condensed aromatic rings in their molecules that carry no substituents or hetero atoms [70, 71]. They consist of three and more aromatic rings containing only carbon and hydrogen. The polycyclic aromatic hydrocarbons are generated by the combustion process of any carbon-containing material, in particular during the thermal decomposition and incomplete combustion of coke and coal, as well as liquid carbonaceous fuels.

Polychlorinated dibenzo-p-dioxins (PCDDs)

Dioxins is a general name for 210 chemicals of two groups technically known as polychlorinated dibenzo-p-dioxins (PCDDs) and polychlorinated dibenzofurans (PCDFs) [69]. PCDDs and PCDFs are the aromatic compounds (hydrocarbons) formed by two benzene rings, which are in PCDDs connected by two oxygen atoms, and in PCDFs by one oxygen atom [70, 73]. Molecules of PCDDs and PCDFs are not very volatile, and when adsorbed on particles generated during combustion, they are highly thermally and chemically stable in the environment. They can decompose only at temperatures above 1000 °C. In this context, it should be mentioned that PCDDs/PCDFs were found not only in flue gases but were also found in solid waste from all combustion processes, in bottom ashes, slag and fly ash. The presence of PCDDs/PCDFs in the environment is considered a significant environmental problem due to the high toxicity of some representatives of this group of substances. Due to the high stability, PCDDs/PCDFs remain in the environment for a long time.

Polychlorinated biphenyls

PCBs are organic compounds where hydrogen atoms on a biphenyl skeleton (a biphenyl hydrocarbon – $C_{12}H_{10}$) are replaced with chlorine atoms to a different extent. PCBs are also produced as unintended by-products in metallurgy, for example, in the combustion of carbon-based waste.

Emissions in the metallurgy of iron and steel

In the following text, typical model examples for emissions of four technological operations in the iron and steel production are given.

For a sinter plant, it is typical that the amount and types of contaminants are very diverse. This relates to the nature of the production, when in addition to the primary aggloores and concentrates, a large number of secondary ferriferous materials is also processed. These materials include various carbon compounds as well. The resulting agglomeration gas contains material particles (heavy metals, particularly iron compounds), but also other compounds (particularly containing zinc and lead), alkali chlorides, oxides of sulfur, oxides of nitrogen, HCl, HF, hydrocarbons, CO and CO_2. The agglomeration gas also contains trace amounts of polycyclic aromatic hydrocarbons (PAHs) and aromatic halogen compounds, such as polychlorinated dioxins and furans (PCDD/F), and polychlorinated biphenyls (PCBs) [70–76]. Although the resulting amounts of PCDD/F per unit volume are relatively low in the process of agglomeration, in the long term, they can cause serious health problems. The overall mechanism of PCDD/F emissions in the agglomeration process is complex, but many authors connect this to the properties of deposited carbon particles (soot), which can be dispersed in the flue gas [74, 76]. In addition to the emissions of waste gas, dust emissions from handling, crushing, screening and transportation of materials for sinter production (including carbonaceous material) are also generated by the sintering process. These secondary dust emissions can be reduced by optimising the handling and treatment processes and the introduction of secondary suction systems.

Recycling of waste gas from the sintering strand can significantly reduce the amount of waste gas (about 28 %) and limit the emissions of pollutants (approximately by 20–30 %). At the same time, this process reduces the consumption of solid fuel (coke powder) by approximately 7–10 kg/t of agglomerate. Agglomeration with emissions optimisation (EOS) was developed primarily to reduce the flow of waste gas, as well as the mass concentration of particulate emissions and PCDD/Fs. Additional separation equipment for further waste gas treatment before discharge into the atmosphere would process smaller volumes with potential savings of financial and operational costs.

In *Tab. 27*, the basic emissions that arise in the production of iron ore agglomerate are given.

Tab. 27 Emissions resulting from production of Fe agglomerate [19]

Parameter	SI unit	Min	Max
Agglomeration gas	Nm^3/t of agglomerate	1500	2500
Dust	kg/t of agglomerate	0.4	15
PM_{10}	g/t of agglomerate	66	177
NO_x	g/t of agglomerate	300	1030
SO_2	g/t of agglomerate	219	970
CO_2	kg/t of agglomerate	160	360
CO	kg/t of agglomerate	9	38
Methane (CH_4)	g/t of agglomerate	35	400
VOC	g/t of agglomerate	1.5	260
PAH	mg/t of agglomerate	0.2	590
PCCD	µg/t of agglomerate	0.15	16

Legend:
VOC – volatile organic compounds
PAH – polycyclic aromatic hydrocarbons
PCCD – polychlorinated dibenzo-p-dioxins

It is peculiar to the agglomeration process that the flue gas collects two types of particulate matter (PM) – crude and fine dust – in its course. The composition of coarse dust is related to the composition of agglomeration raw materials, and it can be separated by electrostatic precipitators (ESP) with high efficiency. On the other hand, fine dust is formed by alkali, chlorides and oxides of zinc and lead, which are formed by chemical reactions during the sintering process. These oxides and chlorides of zinc and lead have a high specific electrical resistance, and thus form an insulation layer on the electrodes of the ESP. This layer causes major problems in de-dusting, reducing the efficiency of electrostatic precipitators. It is possible to achieve the concentration of dust particles below 30 mg/Nm^3 by fabric filters.

Carbonaceous materials can be used as emission absorbents. In sinter plants abroad, various sources of carbon are continuously tested to absorb different kinds of dangerous particles (e.g. heavy metals and PCCD) contained in the waste gases from agglomeration equipment. These include, for example, activated carbon, special brown coal coke, charcoal, and specially prepared carbon-based adsorbents.

The blast furnace production of pig iron is characterised by the use of various carbonaceous reducing agents: carbon (or hydrocarbons) in the form of coke, coal, oil, natural gas, or currently in some cases plastics. BF gas comprises approximately 20–28 % of CO, 17–25 % of CO_2, 1–5 % of hydrogen, 50–55 % of N_2, sulfur oxides, cyanide compounds, polycyclic aromatic hydrocarbons, polychlorinated dibenzo-p-dioxins and a large amount of dust from the charge. During the two-stage treatment of blast furnace gas, dust and also compounds that bind to it are removed with high efficiency, as well as most of the heavy metals and PAHs. The second step of blast furnace gas treatment is the wet separation in the scrubbers, which is the technology very frequently used in the EU [18, 19]. The output from the scrubber is the contaminated wastewater, which contains suspended particles (e.g. carbon and heavy metals), cyanide compounds, nitrogen compounds etc.

CO_2 emissions from the production of pig iron are highly dependent on the type and amount of reducing agents (e.g. coke, coal, oil, natural gas, etc.) used in blast furnaces. For this reason, iron and steel industry have implemented a number of measures for an overall reduction of greenhouse gas emissions, especially CO_2. The technology of production of pig iron is now so optimised that the need for reducing agents is close to the minimum stoichiometric requirement. Energy consumption is continuously reduced by introducing energy efficient equipment into the process and by increasing the efficiency of energy conversion devices. A direct injection of reducing agents through the tuyere is included among the existing technologies that intensify BF process and reduce the emission load. It means that a part of coke is replaced with another source of carbon and hydrocarbons. These hydrocarbons may be in the form of heavy fuel oil, oil residues, particulate or powdered coal, natural gas, or waste plastics. This technology directly reduces coke consumption, overall pollution and reduces the energy requirements. For example, using an oxygen-oil technology, the amount of injected oil is doubled. Coke consumption and CO_2 emissions can be reduced at the same time. The oil is made up of carbon and hydrogen, and substitutes coke in the 1:1.2 ratio (1 kg of oil substitutes 1.2 kg of coke). When using the oxygen-oil technology, the amount of oil is doubled to 130 kg/t of the melt. The coke savings amount to about 15 kg/t of melt and CO_2 emissions are decreased by about 50 kg/t of the melt.

Studies of recycling blast furnace gas are based on a similar idea as the recycling of a part of waste gas from agglomeration strand. Within ULCOS research projects, blast furnace gas is recycled back into the blast furnace in the laboratory and pilot conditions. The technology would utilise pure oxygen (O_2), and carbon monoxide gas (CO) would be re-injected. In the second purification step of BF gas, captured CO_2 is compressed and prepared for the deposition in the geological formations, such as oil fields or natural gas fields, unextractable coal beds, mineral carbonates, or is used for industrial purposes. The technology is ready for use on an industrial scale. In *Tab. 28*, basic emissions resulting from the production of pig iron are listed.

Tab. 28 Emissions from production of pig iron [19]

Parameter	SI unit	Min	Max
Blast furnace gas	Nm^3/t p.i.	1200	2000
Dust	kg/t p.i.	7	40
NO_x	g/t p.i.	30	120
SO_2	g/t p.i.	20	230
CO_2	kg/t p.i.	400	900
CO	kg/t p.i.	300	700
Hydrocarbons	g/t p.i.	130	330
Hydrogen	kg/t p.i.	1.0	7.5
PCCD	µg/t p.i.	0.001	0.004

During the production of steel in the oxygen converter (BOF), oxygen is blown, and converter gas is the product of chemical reactions in the gaseous form. The process of steel production in oxygen converter is primarily a source of dust, solid waste (slag) and wastewater. Converter gas consists mainly of carbon monoxide (CO), large number of particles (mainly comprising the oxides of metals, including heavy metals), and relatively small amount of sulfur (SO_2) and nitrogen oxides (NO_X). In addition, very small amounts of PCDD/F and PAH are emitted as well. Depending on the quality of steel scrap, other organic pollutants, such as chlorobenzenes and PCBs may emerge in the emissions from the converter due to the thermal decomposition of organic materials (oils, paints, lubricants and plastics). Converter gas temperature is about 1200 °C. The flow rate is approximately 50–100 Nm^3/t of steel. This gas contains about 70 to 80 % of carbon monoxide (CO) at the output of the converter, and its calorific value is approximately 8.8 MJ/Nm^3. In *Tab. 29*, the basic emissions occurring in the production of steel in the oxygen converter are given.

The gases that are produced during oxygen blowing (i.e. converter gas) contain large amounts of carbon monoxide. In many steel mills, measures have been taken for the recuperation of converter gas and its use as an energy source. Open burning systems introduce air into the converter flue gas duct and cause burning of carbon monoxide. In the case of suppressed (partial) combustion, air oxygen is not allowed to enter the flue gas duct and carbon monoxide burn-up is thus prevented. Flue gas, rich in CO, can be cleaned and stored for subsequent use as fuel. The use of converter gas in conjunction with blast furnace and coke oven gas (i.e. gaseous products of three metallurgical technologies) brings significant benefits by enabling compensation of large quantities of primary energy sources such as natural gas. The selected type of recovery (partial or complete combustion) affects emissions. When a partial (or suppressed) combustion is used, the lower concentration of particles (5–10 mg/Nm^3) can be achieved.

In terms of reducing the emission load of steel production on the environment, there are currently interesting solutions in the field of net shape steel casting. This technology represents the continuous casting of steel in combination with the direct hot rolling, cooling and belt winding without continuous heating in the oven. It can be used to manufacture thin slabs with a thickness up to 15 mm.

Tab. 29 Emissions resulting from production of steel in BOF [19]

Parameter	SI unit	Min	Max
Converter gas	Nm^3/t steel	500	1000
Dust	kg/t steel	12	23
NO$_x$	g/t steel	5	20
SO$_2$	g/t steel	0.4	5.5
CO$_2$	kg/t steel	11	140
CO	kg/t steel	7	16
PAH	mg/t steel	0.08	0.16
PCCD	μg/t steel	0.001	0.110

Electric arc furnaces (EAF) have nowadays an important place in the concept of the modern steel industry, and a further increase in steel production is expected with this technology in the future. During steel production in an electric arc furnace, an off-gas containing carbon dioxide (CO_2), carbon monoxide (CO), large quantities of solid particles (mainly metal oxides, including heavy metals), relatively small amounts of sulfur oxides (SO_2) and oxides of nitrogen (NO_X) are released. In addition, chlorobenzenes, organic compounds, PCDD/F and PAH are emitted as well. Depending on the quality of the steel scrap used, gas emissions from EAF may contain other organic pollutants such as PCBs and chlorobenzenes due to thermal decomposition of organic materials (oils, paints, lubricants or plastics). In the case of the bituminous coal (anthracite) use, some compounds such as benzene may evaporate before ignition. *Tab. 30* shows the basic emissions that arise from steel production in the electric arc furnace.

In recent years, more and more new EAFs have been fitted with a waste gas preheating system to recover energy. Preheating of the scrap (at temperatures of about 800 to 1000 °C) is successfully used mainly in countries with high electricity costs, e.g. in Japan. This pre-heating is carried out either in scrap bins or a loading shaft (shaft furnace) connected to the EAF, or in a specially designed scrap transport system that allows for continuous feed during the melting process (e.g. CONSTEEL and ESC). In some cases, additional carbonaceous fossil fuels (e.g. natural gas) are also added to the preheating process. Preheating of scrap may be the cause of the production of a large number of organic pollutants due to the possible presence of organic substances in steel scrap that are incinerated during preheating. The result may be increased emissions of volatile substances and PCDD/F. In this case, the waste gases require further treatment and subsequent afterburning.

Tab. 30 Emissions arising from production of steel in EAF [19]

Parameter	SI unit	Min	Max
EAF gas	Nm^3/t steel	200	1200
Dust	kg/t steel	5	30
NO_x	g/t steel	120	240
SO_2	g/t steel	24	130
CO_2	kg/t steel	2	50
CO	kg/t steel	0.7	4
PAH	mg/t steel	3.5	71
PCCD	µg/t steel	0.07	9

In combination with the advanced treatment of waste gas, preheating of scrap steel plays an important role in optimising the production of steel in electric arc furnaces, not only for increased productivity but also to minimise emissions. As a side effect of scrap preheating, dust emissions are reduced by about 20 % because the waste gas has to pass through scrap that acts as a filter. This increases the zinc content in the dust, which can be recycled more efficiently, and this improves the environmental performance of steel production in an EAF.

11. Alternative fuels

Metallurgy is one of the sectors that contribute to raising standards of living but, on the other hand, it is also a source of environmental pollution. Such sources include mainly non-renewable energy sources – fossil fuels based on coal and hydrocarbons. The role of searching for new additional energy carriers for the metallurgical industry in Slovakia is highly relevant. By introducing suitable high-efficiency technologies that will use alternative energy sources, it is feasible to utilise otherwise unusable waste, with minimum impact on the environment. However, such technologies should also be sufficiently reliable. One of the important goals of metallurgical companies and research institutes worldwide is to examine the partial replacement of fossil fuels with materials from biomass. This goal could be met most effectively using biomass from local sources.

Fig. 26 illustrates the basic classification of alternative fuels or biofuels by state based on renewable biomass. Before dealing with solid biomass (which is still the most interesting for the metallurgical industry), let us briefly characterise some important alternative fuels in liquid and gaseous state, as well as solid fuels, which are also included in the group of alternative fuels even though they originate from fossil fuels, *Fig. 27*.

Alcoholic biofuels

The most utilised alcoholic biofuels are bioethanol and bio-methanol. Bioethanol (ethyl alcohol) or bio-methanol (methyl alcohol) is the fuel produced by alcoholic fermentation. It is made of organic residues or from specially produced crops. The main raw materials for the production of bioethanol are those with high sugar content, starch and lignocellulose (e.g. sugar cane, sugar beet, potatoes, fruits, etc.). The advantage of using alcoholic biofuels is that lower amount of harmful substances is produced during their combustion. These fuels have a simpler structure than gasoline or diesel, burn better, and the entire burning process leads to the formation of fewer unburnt residues. Alcoholic biofuels are, so far, not used in metallurgical processes of iron production and steel production.

Biodiesel

The biodiesel label has been introduced for oily acid methyl esters. The raw material for biodiesel production is vegetable oil (e.g. rapeseed oil) obtained from oily plants or animal fats (e.g. fish oil, beef fat). Biodiesel is not used in the metallurgical processes of iron and steel production either.

Biogas

Biological decomposition of organic substances that are the basis of biomass is a complex multistage process, where some important chemical reactions are taking place. The action of microorganisms in anaerobic conditions generates biogas, which is mostly composed of two gaseous components – methane (CH_4) and carbon dioxide (CO_2). The methane content is about 40–80 %, and the carbon dioxide content is about 20–60 %. Biogas also contains other gaseous components, hydrogen (H_2), hydrogen sulfide (H_2S) and water vapour ($H_2O_{(g)}$). The process of biological decomposition of organic matter can also take place in solid biomass, which is used for metallurgical purposes. It involves chemical reactions resulting in the decomposition of glucose ($C_6H_{12}O_6$), acetic acid (CH_3COOH) and methanol (CH_3COOH).

$$C_6H_{12}O_6 \rightarrow 3\ CH_4 + 3\ CO_2 \qquad \Delta H_{298} = -141.827\ kJ \qquad (31)$$

$$CH_3COOH \rightarrow CH_4 + CO_2 \qquad \Delta H_{298} = 17.657\ kJ \qquad (32)$$

$$4\ CH_3OH \rightarrow 3\ CH_4 + CO_2 + 2\ H_2O \qquad \Delta H_{298} = -234.176\ kJ \qquad (33)$$

Biogas is, as yet, not used in the metallurgical process for the production of iron and steel.

Hydrogen as an alternative energy source

Although hydrogen is not a carbonaceous fuel, it can be produced from alternative carbonaceous fuels. Hydrogen is widespread in nature mainly in the water molecule (H_2O). Its recovery is possible by the electrochemical decomposition of water, which is called water electrolysis. However, hydrogen is not the primary source of energy. It is only an energy carrier. Water decomposition is carried out according to the chemical reaction:

$$H_2O_{(l)} \rightarrow H_{2\ (g)} + \tfrac{1}{2}\ O_{2\ (g)} \qquad \Delta H_{298} = 571.660\ kJ \qquad (34)$$

Benefits of using hydrogen:

- Burning of hydrogen produces only trace amounts of hydrocarbons, carbon monoxide or carbon dioxide, thus eliminating emissions that cause environmental pollution and global climate change.

- Higher energy efficiency.

- It can be produced from various energy sources.

Disadvantages of using hydrogen:

- To obtain hydrogen, it is necessary to introduce the input energy.

- Explosiveness of hydrogen.

- Storage of hydrogen.

- High production costs of hydrogen.

Hydrogen is not directly used in metallurgical processes of iron and steel production. Nevertheless, there are hydrocarbon fuels releasing hydrogen, which is used for combustion or as a reducing agent (e.g. coal gasification gas, gas from pyrolysis of plastics, etc.).

Plastics

Plastics are organic substances that are made from crude oil and become a source of energy as waste. Every year, up to 100 000 tonnes of plastic waste are produced in Slovakia, which is about 7 % of the total amount of municipal waste. The energy potential of plastics is about 18–22 MJ/kg. The potential possibilities of preparing liquid or gaseous fuel from plastic waste and using it as a replacement for heavy fuel oil or natural gas have been known for a long time. However, the implementation has not yet gone beyond the pilot level. Abroad, plastics are used in some metallurgical technologies. Probably the most advanced and sophisticated technology is in Japan, where the plastics are injected into the tuyere manifold of BF after the adjustment of granulometric composition.

Rubber

In the energy recovery of rubber, its high energy potential (about 30–34 MJ/kg) is utilised by high-temperature oxidation or pyrolysis. Disposed worn tires represent a relatively large amount of material with high energy content. The composition and calorific value of rubber are comparable to the high-grade types of bituminous coal. Combustion is carried out in special furnaces. In Slovakia, approximately 24 000 tonnes of waste tires are generated in Slovakia annually. Some rubber pyrolysis products are also experimentally tested in metallurgical processes (e.g. as a reducing gas for direct iron production (Mexico – the Midrex process).

11.1. Biomass

Biomass has the highest usable potential of all renewable energy sources for metallurgical use. There are high hopes regarding biomass (mainly dendromass and phytomass) intended for energy and metallurgical becoming an alternative renewable energy source and gradually replacing a part of the non-renewable classical fossil fuels. Although the increased use of alternative fuels based on organic materials makes it possible to reduce the consumption of fossil fuels and hence the production of gaseous pollutants, there has not been much progress made in this respect. The main obstacles include:

• distorted prices of primary energy sources,

• absence of distribution network for new types of energy carriers,

• high investment costs for new technologies,

• absence or deficit of a comprehensive analysis of the use of alternative fuels in new technologies,

• large regional differences in the sustainable availability of alternative fuels, which depend on the real support of EU countries – and there are big differences between different regions.

In many European countries, there is a legislation in place to assess the quality of alternative solid fuels. This legislation is a basic communication tool for manufacturers, traders and consumers of biofuels. For alternative fuel consumers, it is important that these fuels have the highest quality and homogeneity. Recently, we have encountered the term of standardisation of alternative solid fuels [77, 78]. Mainly wood biomass is standardised – namely by means of pelletisation, (*Fig. 101*). However, there are requirements to mould also other types of waste besides wood biomass. The process of refining ensures the homogenization of the fuel composition, the homogenization of its size, its moisture, the adjustment of its composition and other physical-mechanical properties. This process improves the quality of alternative solid fuels.

Fig. 101 Compressed biomass – standardisation of alternative solid fuels

The effective change of fuel type from fossil fuel to an alternative biofuel must be preceded by comparative analysis. It is very difficult to carry out an appropriate analysis that accurately characterises the substitution of primary energy sources (mainly fossil) by alternative fuels. The purpose of such analysis is to evaluate the state and the amount of biomass with a real energy equivalent. The qualitative analysis compares the technological, environmental and economic parameters of alternative biofuels with classic fossil fuels. The reserves of raw biomass presently exceed their use for energy purposes and industrial use several times. Nevertheless, the future trend should be the use of mainly waste biomass. This publication focuses primarily on the impact of biomass on technological, qualitative and ecological parameters of some important processes in metallurgy (e.g. impact of biomass on the agglomeration process).

11.1.1. Classification of biomass

There are several definitions specifying biomass, but they all start from the same basis that biomass is a mass of organic origin. Biomass means biodegradable fractions of products, waste and residues from agriculture, forestry and related industries, as well as biodegradable fractions of industrial and municipal waste. Another definition refers to biomass as a substance of biological origin, which includes plant biomass grown in soil and water, animal biomass, the production of animal origin and organic waste, *Fig. 102*.

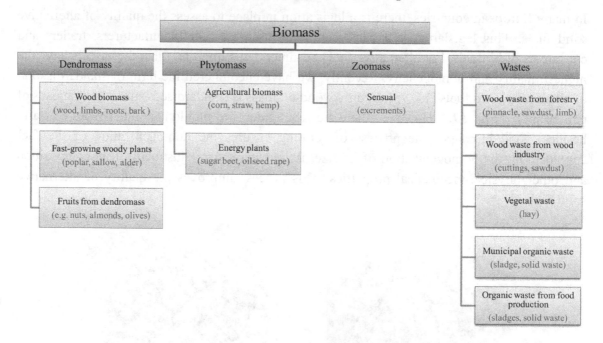

Fig. 102 Basic classification of biomass

The main producers of biomass raw materials are, above all, those sectors of the economy that can produce biomass quantities usable on an industrial scale. It is mostly the agricultural and food industry whose by-products find a variety of uses, but they are increasingly becoming potential sources for the production of biofuels as well. The forestry and woodworking industry, where the raw material differentiation for material use and energy recovery has been taking place lately, is not an exception. The cultivation of fast-growing trees at plantations is a sort of an intersection of agriculture and forestry. The basic classification of wood and plant biomass is shown in *Tab. 31*. It is clear that the sources of wood and plant biomass are plentiful. Utilization of these resources for the energy sector and the industry is, however, strictly limited because of their physical-mechanical and chemical properties and already mentioned inhomogeneity.

Tab. 31 Classification of wood and plant biomass

Dendromass		
Forest and plantation wood	**Wood processing industry**	**Used wood**
whole trees and shrubs	chemically untreated wood residue	chemically treated
logs	chemically treated wood residue	chemically untreated
residues after harvest	fibrous waste from cellulose and paper industry	
stumps		
bark		
wood biomass from landscape care		
Phytomass		
Agricultural and garden plants	**Industry processing plants, by-products and residues**	**Fruit biomass**
cereals	chemically treated residue	industry processing fruit, by-products and residues
grasses	chemically untreated residue	orchard and garden fruits
oilseed		
root crops		
legumes		
flowers		
herbal biomass from landscape care		

11.1.2. Properties of biomass

The basic specification of carbonaceous fuel has been described in chapter 6. "Carbonaceous fuels, their properties and testing". The biomass also belongs to carbonaceous fuels, so it is clear that the same methodology as for any fossil fuel will be used for its assessment. Biomass differs according to its origin, and its physicochemical properties are closely related to the formation of its material. In the following text, the most important biomass properties that are associated with its use within the thermal combustion processes will be specified.

The basic organic material of the wood biomass consists of cellulose, hemicellulose and lignin. The biomass structure affects its properties and the thermal processes of its processing. Although the chemical composition of biomass differs between plant species, plants contain on average about 20–40 % of lignin and 60–80 % of carbohydrates or sugars that consist of many sugar molecules in long polymer chains. The two most important carbohydrate components are cellulose a hemicellulose. Nature uses long cellulosic polymers to build fibres that give plants the necessary strength. The lignin component acts as a binding substance that holds cellulosic fibres together. In **Tab. 32**, there is the structural composition of some biomass types, which means that, for example, nuts and olives have the majority share of lignin and almonds, and sunflowers have the majority share of cellulose.

Tab. 32 Average results of biomass sample structures

Composition	Share of basic structural components in biomass samples [%]				
	hazelnut	walnut	almond	sunflower	olive
hemicellulose	30.4	22.1	28.9	34.6	23.6
cellulose	26.8	25.6	50.7	48.4	24.0
lignin	42.9	52.3	20.4	17.0	48.4

The basic elements of biomass matter are carbon, hydrogen, oxygen and nitrogen. Plant and wood biomass contains the highest amount of hydrogen and especially oxygen of all existing solids carbonaceous fuels. With the increasing ratio of H/C atoms, as well as the O/C ratio in biomass, its calorific value decreases, see **Fig. 30** (Van Krevelen diagram). The increase in calorific value thus directly depends on the increase of carbon content and the reduction of oxygen and hydrogen content. The most important component in obtaining thermal energy from carbonaceous fuel is carbon. Its chemical and physical properties determine the use of biomass as a fuel. The carbonaceous fuel composition is illustrated in **Fig. 103**. The basic composition of carbonaceous fuel is supplemented by the specification of differences between biomass and coke composition. When we look at the biomass composition (**Fig. 103b**), we discover that it contains a significantly lower fixed carbon content, and on the other hand, has a significantly higher content of volatile combustible. Since the biomass has a higher oxygen and hydrogen content, a part of volatile combustible will be bound to so-called constitution water, and carbon dioxide will be released together with the volatile combustible. These two components will reduce the calorific value of biomass.

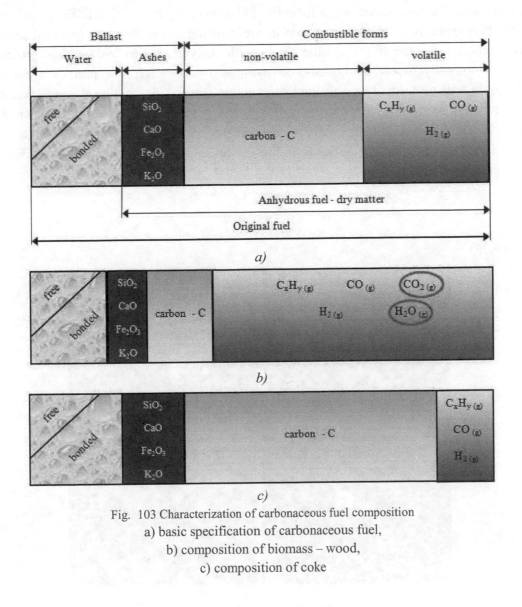

Fig. 103 Characterization of carbonaceous fuel composition
a) basic specification of carbonaceous fuel,
b) composition of biomass – wood,
c) composition of coke

One of the factors affecting the quality of biomass is the water content that has a direct effect on the calorific value. The water content in biomass is very variable and depends on the type of biomass, the way of storage and processing.

In *Fig. 103*, wood biomass having a lower water content than coke is shown in the comparison of biomass and coke composition. It is a model comparison. In reality, wood may have a higher content of water than solid fossil fuels (coal and coke). Fresh (raw) biomass usually has high content of water, which has a large evaporation heat. Prior to heat treatment (e.g. combustion), the biomass needs to be dried. It is generally advisable to reduce the moisture below 30 %, while the optimum is less than 20 %. To press briquettes or pellets, the raw material should be dried to even lower water content (below 15 %).

Fig. 104 shows the microstructure of fuels from biomass [79]. In addition to fixed carbon, the hydrocarbon group ($C_XH_YO_Z$) can be seen in the structure of raw biomass (pressed sawdust) as well, which gives rise to the volatile combustible. Oxygen in this hydrocarbon group will be bound not only to carbon but also to hydrogen, and therefore a part of the volatile combustible will be formed by so-called constitution water. Constitution water is converted into water vapour in thermal processing of biomass. Compared to coke, charcoal and dry straw have a higher porosity, especially micropores that spread individually. In coke, there are mainly larger pores that are formed in the processes of coal carbonisation during the production of coke.

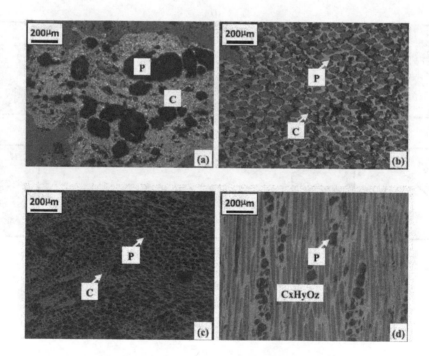

Fig. 104 Solid fuel microstructures [79]
a) coke powder, b) charcoal, c) dry straw, d) pressed sawdust

Legend: C – carbon, P – pores, $C_xH_yO_z$ – hydrocarbon group

Tab. 33 gives the water content and calorific value of some biomass types [1]. It is clear that dendromass typically has a higher water content than phytomass. The water content significantly affects the calorific value of the fuel not only by reducing the dry matter content but also by the said consumption of energy for evaporation.

Tab. 33 Calorific value and water content of some biomass types [1]

Solid fuels	Calorific value (CV) [MJ.kg^{-1}]	Water content (W) [%]
Pine wood	18.2	20
Spruce wood	15.3	18
Beech wood	15.5	20
Oak wood	15.9	18
Willow wood	16.9	22
Poplar wood	12.9	19
Birch wood	15.0	20
Wood chips	12.8	30
Cereal hay	15.9	10
Corn hay	14.4	11
Rape straw	16.2	10

In *Tab. 34*, the chemical analysis (primary and basic) of some biomass fuels is shown. *Tab. 35* gives the chemical composition of ash in some biomass fuels. Already discussed facts that the original biomass has a significantly lower fixed carbon content and ash, and significantly higher volatile combustible content than coke (or coke powder) result from both tables. Charcoal represents the refined form of fuel from biomass when low-ash wood is used for its production. The pyrolysis process increases the fixed carbon content, reducing the amount of volatile combustible in charcoal.

Tab. 34 Chemical analysis of biomass fuels

Type of fuel	Proximate – approximate analysis [wt%]				Ultimate – element analysis [wt%]					
	H$_2$O (W)	Ash (A)	Combustible volatile matter (VM)	Fixed carbon (C$_{FIX}$)	Ash (A)	C	H	O	N	S
Sunflower husks	9.3	3.20	75.50	12.0	3.20	46.8	6.10	43.10	0.70	0.10
Hazelnut shells	11.0	1.09	67.80	20.0	1.09	55.1	6.12	37.23	0.42	0.04
Almond shells	9.2	0.49	81.37	8.8	0.49	48.8	6.41	44.09	0.18	0.03
Rape	8.1	4.50	83.18	3.6	4.50	54.1	8.58	28.08	4.14	0.60
Wood chips	5.2	1.90	76.10	16.8	1.90	55.6	5.80	36.39	0.21	0.10
Charcoal	2.2	1.80	6.40	89.6	1.80	92.4	1.40	3.90	0.40	0.05
Coke powder	5.5	12.1	1.50	80.9	12.10	85.4	0.30	0.50	1.30	0.50

Tab. 35 Chemical composition of ash from biomass fuel

Type of fuel	Ash content [%]	Chemical composition of ash* [%]						
		SiO_2	Al_2O_3	Fe_2O_3	CaO	MgO	K_2O	P_2O_5
Sunflower husks	3.2	4.49	0.75	4.79	19.06	12.02	30.95	8.92
Hazelnut shells	1.1	20.00	1.20	6.43	27.70	6.80	15.90	9.70
Olive pits	9.5	18.70	4.08	3.28	28.10	7.02	29.10	4.68
Sugarcane	4.3	55.00	8.50	5.50	12.00	4.79	7.65	3.43
Charcoal	4.5	3.20	0.98	1.10	61.70	4.80	2.20	0.95
Coke powder	12.5	34.70	21.10	27.20	6.80	2.80	1.60	0.64

* Calculated to ash content

Besides basic chemical elements, biomass also includes elements that affect the production of harmful substances in its use (e.g. combustion). These include sulfur and chlorine. The biomass contains yet a lot of other so-called trace inorganic elements that affect its use and production of harmful substances, deposits, etc. These include lead, potassium, sodium, calcium, magnesium, silicon, manganese, iron, copper, nickel, zinc, etc. These elements are mostly occurring in oxidic form in the ash of the biomass.

Characteristics of ash (chemical composition and fusibility) of different types of biomass are also shown in *Tab. 36* [80]. The methodology of preparing such ash from biomass is very specific, and it also requires knowledge of the ignition temperatures for different types of biomass (generally around 220–500 °C). The ash, the characteristics of which are given in *Tab. 35*, was prepared by heating the individual biomass in an electric resistance furnace at 550 °C. The heating took 4 hours in oxidation atmosphere until this temperature was reached. The ash obtained was sifted through a 0.25 mm sieve after the experiment. Fusibility was determined by the high-temperature observation of the pellet ash (3 mm = diameter and height) during heating from 25 °C to 1400 °C. The heating rate was 60 °C/min to 600 °C, and from this temperature 10 °C/min up to 1400 °C. The device used was an optical high-temperature LEICA microscope. The characteristic measured temperatures were determined according to the relevant standards (DIN 51730 – 1998 and DIN 51730 – 1994). It follows from the analysis that plant biomass has lower melting and creep temperatures, which enhances the probability of ash sintering of this biomass during combustion compared to dendromass.

Tab. 36 Characteristics of biomass ash [80]

Biomass	Ash [%]	Chemical analysis of ash [%]					Fusibility [°C]			
		CaO	K_2O	MgO	Na_2O	SiO_2	PD*	S*	H*	K*
pine	3.1	13	7.9	4.5	1.90	52	1190	1200	1220	1280
eucalyptus	4.3	18	8.7	4.2	1.90	41	1160	1170	1190	1230
cork	4.5	35	5.1	1.4	0.70	20	1190	1200	1220	1280
poplar	3.4	33	18.0	3.7	0.14	2.8	> 1400	> 1400	> 1400	> 1400
sunflower	9.7	17	24.0	3.8	0.55	11	740	nd	1360	1390
wheat hay	5.9	8.1	18.0	2.4	0.22	44	850	1040	1120	1320
rice hay	13.1	8.9	16.0	3.5	2.80	51	860	980	1100	1220
thistle	14.1	14	15.0	2.5	9.20	13	640	660	1150	1150
almond shells	1.0	15	22.0	1.5	0.30	4.6	750	770	nd	> 1400
olive pits	3.4	4.4	27.0	1.7	0.52	24	1030	nd	1090	1160

* ID – initial deformation, S – sphere (start of melting), H – hemisphere (melting),
  L – liquid (flow), nd – not determined

The content of ash in biomass is significantly lower than in fossil fuels (coal and coke). The amount of ash in biomass is typically from 0.2 to 7 % (it does not exclude even higher ash content). In phytomass, there are usually higher values of the said interval, while they are lower in dendromass. Biomass is characterised not only by lower levels of ash but also by its properties. Ash from biomass has lower melting temperatures due to higher amounts of potassium, nitrogen and chlorine. It is apparent based on *Tab. 35* that all types of biomass ash are higher in K_2O content than carbonaceous fossil fuels. Mainly chlorides and potassium oxides in a ternary system with CaO and SiO_2 significantly reduce the melting temperature range, *Fig. 105*. Melting temperatures of the pure complex compounds $2K_2O – CaO – 6SiO_2$, $2K_2O – CaO – 3SiO_2$, $K_2O – SiO_2$, $K_2O – 2SiO_2$ and $K_2O – 4SiO_2$ are in the temperature range of approximately 900–1200 °C. Plant biomass has a higher alkali content than dendromass. The melting temperatures of plant biomass are about 850–1100 °C. In the case of dendromass and fossil fuels, they are higher (about 1100–1450 °C).

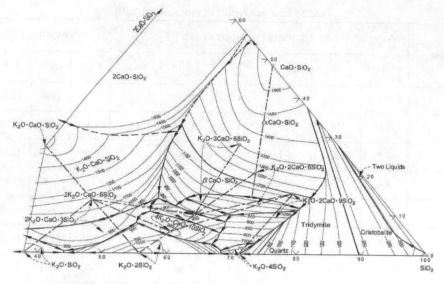

Fig. 105 Phase diagram of $SiO_2 - K_2O - CaO$ system

Fig. 106, 107 show the microstructures of ash produced by combustion of carbonaceous fuels [81]. While ***Fig. 106*** shows the microstructure of ash produced at temperatures of 1100−1248 °C with a minor proportion of plant biomass (30 % rice straw), ***Fig. 107*** shows the microstructure of ash produced at temperatures of 1094−1229 °C with a majority share of wood biomass (80 % sawdust). At comparable temperatures, the ash of the plant biomass is substantially melted, while non-melted original grains can be seen in ash with a share of wood biomass.

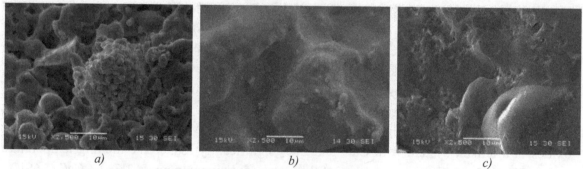

Fig. 106 Microstructure of ash (30 % rice straw + 70 % coal) [81]
a) 1100 °C, b) 1194 °C, c) 1248 °C

Fig. 107 Microstructure of ash (80 % wood sawdust + 20 % coal) [81]
a) 1094 °C, b) 1207 °C, c) 1229 °C

11.1.2.1. Dendromass

The trees represent the highest life form of plants. Unlike other species, a substantial part of their biomass consists of wood. Therefore, this group is called wood biomass or dendromass. Wood biomass is an organic material, with the highest share of cellulose (42 %), hemicellulose (26 %), lignin (25 %) and other ingredients. In general, the specificity of wood and biomass is that it contains the greatest proportion of volatile combustible (75 to 85 %) released during the combustion process. *Fig. 108* shows the weight loss of the hard medium-coalified coal and pine wood during heating in oxidating conditions [82]. The release of volatile combustible and subsequent burning occurs in pine wood in the temperature range of 300–600 °C. In coal, this process occurs at higher temperatures (about 450–900 °C).

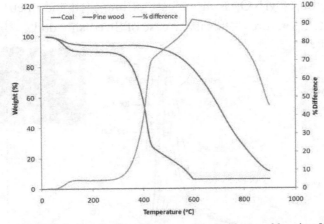

Fig. 108 Weight loss for bituminous coal and pine wood heating [82]

Fig. 109 shows the derivatographic analysis of two species of dendromass (miscanthus and Swedish wood) at different grain sizes [44]. Derivatographic analysis was carried out in inert conditions. During the heating process, release of gaseous components (mostly volatile combustible) occurs in the temperature range of 300–550 °C (miscanthus) and in the temperature range 300–600 °C (Swedish wood). From DTG records, it is clear that the rate of weight loss is highest in miscanthus within the highest grain size class (1180–600 µm) at the temperature of about 480 °C. By contrast, the rate of weight loss is the highest in Swedish wood within the lowest grain size range (53–75 cm) at the temperature of about 420 °C. Within the structure of wood, long fibres (in the case of Swedish wood these fibres were also branched) were identified in both types of dendromass, with so-called tracheids arranged in irregular rows. The structure of biomass, i.e. the main parts of organic matter – cells, has a significant effect on the thermotechnical processes. In these processes, volatile substances are released, and gaseous products of chemical reactions (mainly oxidation) are formed. As a result of the occurrence and the course of these reactions, weight loss occurs in the bulk of the biomass within thermodynamic and kinetic conditions. The speed and temperature interval of the biomass weight loss depends on its reactivity. It is defined by the physicochemical properties and mainly by the structure of the given biomass. Reactivity is primarily associated with oxidative processes, but it can also be perceived in terms of degradation of the matter by other than oxidative processes. In biomass fuels, there is about 60–85 % of the volatile combustible depending on their type. It is, therefore, the majority of the mass that changes into the gaseous form due to the increasing temperature. It has already been said that the structure of raw biomass also includes the hydrocarbon group ($C_XH_YO_Z$), from which the volatile combustible originates. At the temperatures lower than the fuel degradation temperatures of biomass, the hydrocarbon group $C_XH_YO_Z$ is a part of the cellular structure. The conversion of this hydrocarbon group into gaseous components depends on the shape, size, layout and composition of the majority biomass mass cells. Therefore, it is also important to examine the biomass structure and to find the context of its transformations in the thermotechnical processes. The intense release of volatile combustible within the lowest grain sizes of Swedish wood is likely to be related to the structure of this biomass. The release of the hydrocarbon group $C_XH_YO_Z$ is in the case of branched and open fibres with tracheids possible even at low temperatures – this is the case of Swedish wood.

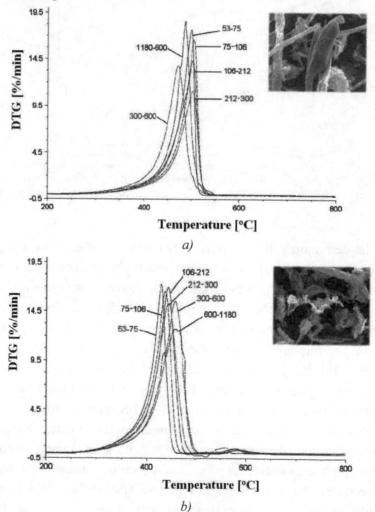

a)

b)

Fig. 109 Records from derivatographic analysis of various types of biomass at different grain sizes (in μm) [44]

a) miscanthus b) Swedish wood

Wood is the oldest used solid fuel in the history of humankind, and it is popular for its favourable properties such as calorific value and especially availability even today. The wood matter, which is not used for construction and other material purposes, gains importance especially in the field of energy production. The options of biomass utilisation are broad, and besides its use in the raw or dried state, the thermochemical treatment is often required. Wood conversion by the pyrolysis process helps improve its qualitative parameters and allows obtaining more refined fuelwood charcoal.

Wood charcoal

Wood charcoal as a solid fuel is widespread throughout the world. Nowadays, it is used mainly in homes for heating. In some countries, it is used in the manufacturing of iron or ferroalloy electrometallurgy as a reducing agent. Annual global production of charcoal reaches a value of over 10 million tonnes (yr. 2014). Charcoal is a solid product of the thermal decomposition of wood biomass, where wood is carbonised at high temperatures without the access of air, *Fig. 110*. The carbon content of charcoal depends on the pyrolysis temperature, which is affected by the physical properties and typically ranges from 80 to 85 %. The volatile content in charcoal is 5–20 %, and ash content, which depends on the content of crude ash in wood material, is approximately 0.2 to 5 %. Weight loss for charcoal at 500 °C was in the range of 20 to 25%, and it was about 30 to 35% at 1000 °C. Properties of charcoal, such as density, reactivity and thermal conductivity, depend namely on the temperatures at which coal was obtained. These properties are subsequently significantly influenced by the structure of a wood material, which changes at different carbonisation temperatures.

Fig. 110 Picture of charcoal sample

The structure of charcoal can be studied using optical methods – e.g. within a microscopic analysis. The charcoal sample preparation before its optical observation depends on the needed form in which the sample is analysed. Raw charcoal (in lump form) can be vacuum cold-cast in Epovac. The preparation of the charcoal sample is performed by grinding it to a fraction under 1 mm and removing the finest dust particles. Microscopic analysis of charcoal has shown that the substance of the material is formed by libriform fibres with relatively small diameter, arranged closely side by side, **Fig. 111**.

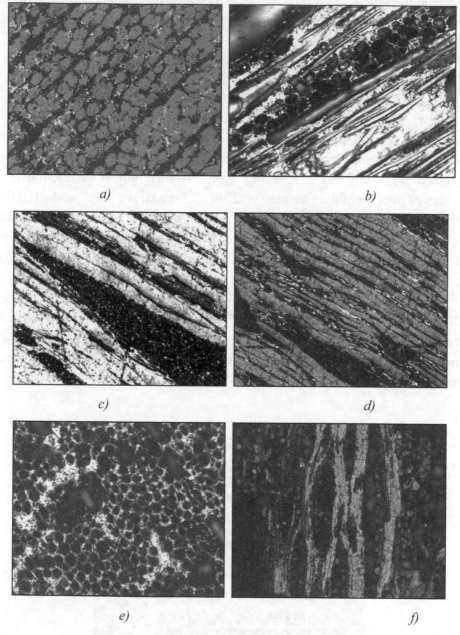

a) *b)*

c) *d)*

e) *f)*

Fig. 111 Optical observation of charcoal samples
a, b, c, d) untreated charcoal – lump, e, f) prepared charcoal – ground

Microscopic structure of charcoal is also shown in **Fig. 112**. Hardwood (oak and beech) consists mostly of libriform fibres that are also retained in charcoal. They are arranged close to one another and have a small diameter. Large fragments of originally circular cross-section are the remains of veins. The structure of soft softwood (pine) is considerably simpler compared to hardwood. In the image (*Fig. 112b*), the long vertical hollow fibres, so-called tracheids, arranged in uniform radial lines can be clearly identified [84]. These elements account for over 90 % of the whole composition. Porosity observed in charcoal from hard and soft wood is 2.9-fold or 3.4-fold greater than in coke. This means substantial enlargement of the reaction surface, especially while maintaining the same particle size of coke and charcoal fuel.

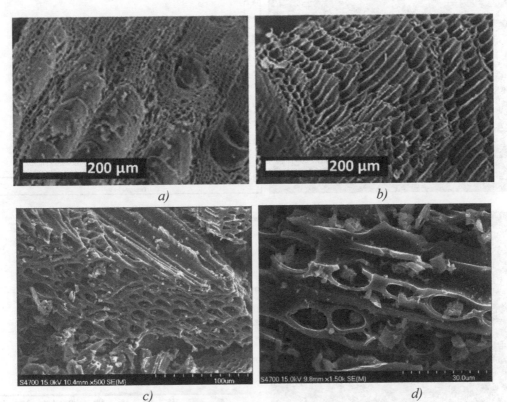

a)   *b)*

c)   *d)*

Fig. 112 Microstructure of charcoal produced from wood [84]
a) oak, b) pine, c, d) beech

Microstructural analysis by scanning electron microscopy (SEM) and energy-dispersive analysis (EDX) can complete the picture of charcoal morphology and confirm the presence of the elements found by chemical analysis. The results of microscopic observation and elemental analysis showed no significant differences. The analysis of the samples confirmed the higher porosity of charcoal. Scanning electron microscopy can detect and specify details of charcoal sample morphology, identification and arrangement of particles observed by optical microscopy. It is possible to determine their elemental composition on the polished surface using X-ray spectroscopy (EDS detector – EDX X-ray analysis). With the increasing magnification and penetration into the structure of the matter, it can be observed that the particles are rather aggregates, comprised of tens to thousands of primary particles. Carbonaceous particles, mineral components and organic compounds are then incorporated in their volume. These identification methods provide unique information and impose certain requirements on sample preparation and interpretation of results.

The apparent size of charcoal conglomerates varies in the range of 100–500 μm. Particle size differences are significant, and the chemical composition of the analysed areas captures mainly mineral components of charcoal, *Fig. 113*.

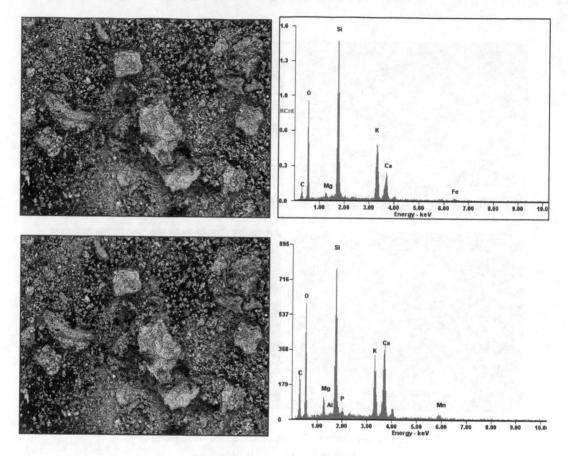

Fig. 113 EDX analysis of charcoal

Wood chips and sawdust

Chips are made from wood waste, e.g. from thinning of growth or branches, *Fig. 114*. The length of wood chips is between 5 and 50 mm, width is 5–30 mm, and thickness is 5–15 mm. The particles of larger dimensions are allowed up to 3 % and the particles of smaller dimensions up to 10 % of chip weight in the natural state of moisture. The optimum relative moisture of wood chips for energy and industrial applications is 30 to 35 %. The produced chips decompose rather quickly due to the activity of living parenchymatic cells, chemical oxidation, hydrolysis of the cellulosic component in an acid medium, and the biological activity of bacteria and fungi, thus reducing their volume. It is therefore important to store this type of biomass in dry conditions and process it quickly.

Fig. 114 Samples of wood chips

Fig. 115 shows the results of derivatographic analysis of wood chips at varying particle sizes [44]. Derivatographic analysis was carried out in inert conditions. During heating, the release of gaseous components (mainly volatile matter) occurs in the temperature range of 350–700 °C. From the DTG results, it is clear that the rate of weight loss of wood chips is the highest in the grain size class of 75–106 mm at about 580 °C. The rates of weight loss in all analysed grain size classes of wood chips are significantly lower than for crude dendromass (e.g. miscanthus and Swedish wood, *Fig. 109*). Within the structure of wood chips, tracheids were identified as a part of fibres that have been layered in the mesh structure. In the case of such structure, the release of the hydrocarbon group $C_XH_YO_Z$ is difficult, the intensity of the release of gas components is lower, and it takes place at higher temperatures.

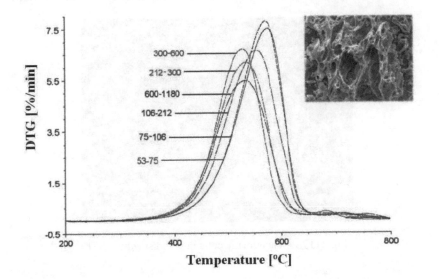

Fig. 115 Record of derivatographic analysis of wood chips at different particle sizes (in μm) [44]

Sawdust consists of minuscule pieces of wood. It is produced as a by-product of sawing or any other machining (e.g. sanding). With regard to the wood biomass, sawdust has the smallest particle size. The size of particles ranges from 0.01 to 2 mm, *Fig. 116*.

a)   *b)*

Fig. 116 Pictures of sawdust samples
a) oak sawdust, b) pine sawdust

Sawdust moisture is very variable and ranges from 10 to 40 %. Due to the activity of living parenchymatic cells and biological activity of bacteria and fungi, sawdust also decomposes relatively quickly. It has a limited storage life, and prompt processing is important. Wood sawdust has a similar structure as the raw dendromass, from which it originates. The most commonly it consists of two basic building elements – tracheids and parenchymatic cells, *Fig. 117*. Wood sawdust generally has low calorific value of about 15 to 18 $MJ.kg^{-1}$.

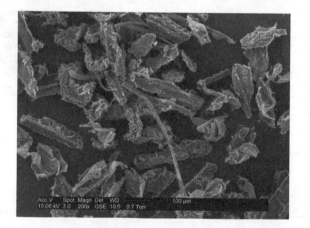

Fig. 117 Parenchymatic cells in wood sawdust [85]

Samples of pine and oak sawdust are usually visually observed without modifying their particle size composition (as opposed to charcoal). In pine softwood sawdust, finely porous regular structure composed of parenchymatic cells and tracheids as the principal building elements was observed, *Fig. 118*. The tracheids as veins are found in woody parts of plants in the form of tubes consisting of elongated cells. They make up 87–95 % of the total volume of wood. They are closed elongated cells with different endings, and their shape, size and thickness of the cell mass are influenced by their function in growing trees. Pine sawdust parenchymatic cells have the form of short prisms or cylinders and constitute 5–10 % of the total volume of wood. In oak sawdust, a substantial part comprised sclerenchymatic cells that form thick-walled libriform fibres. The amount of libriform fibres in wood and especially the thickness of the cell wall affect the density and strength of wood. The absorbed water is stored in the submicroscopic cervices, which is very important in terms of wood hygroscopicity.

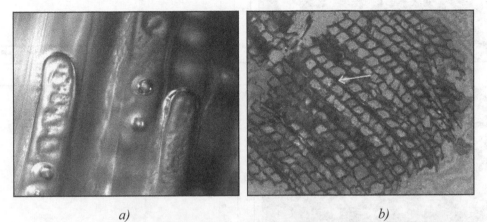

a)   b)

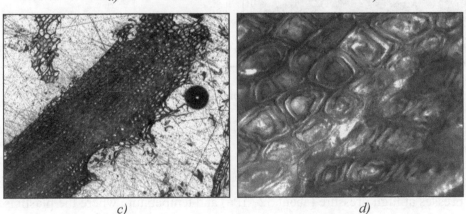

c)   d)

Fig. 118 Optical observation of sawdust sample
a, b) parenchymatic cells and tracheids of pine sawdust,
c, d) sclerenchymatic cells of oak sawdust

The sawdust of pinewood is relatively uniform, and the size of conglomerates is approximately between 50 and 100 μm. The chemical composition of the conglomerates includes C, O, Si, Ca, Al, K, Mg and other elements. The majority of the particles is characterised by a predominant amount of calcium. However, particles with a predominant amount of silicon and aluminium are also present in small amounts, *Fig. 119a*.

The size of conglomerates of oak wood is in the range of 20–70 μm. They are also quite uniform, and the overall chemical composition is very similar to the material of pine sawdust, *Fig. 119b*.

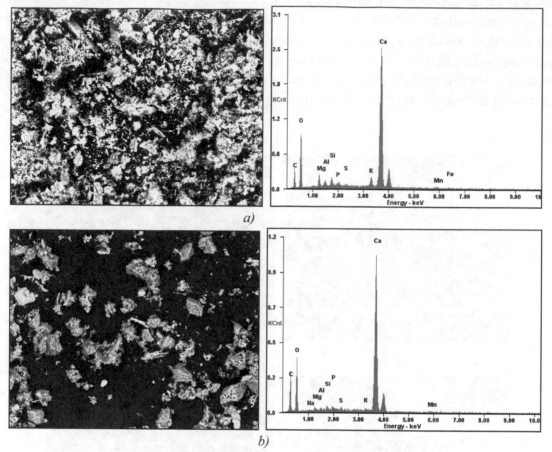

Fig. 119 EDX analysis of pine (a) and oak (b) sawdust

Wood briquettes and pellets

Wood briquettes are manufactured from wood waste of suitable grain size and moisture in briquette presses at high pressure (about 32 MPa) and temperature where the plasticised lignin becomes a binder. By pressing, the density of about 1200 kg.m^{-3} is achieved, which is important to minimise the volume of fuel. Wood briquettes have a higher calorific value than, for example, sawdust, namely up to 20 MJ.kg^{-1}. The briquettes are characterised by a low content of sulfur (about 0.07 %), low ash content (0.5 %), unlimited storage stability, low dust and easy handling. Wood briquettes are diverse in terms of their shape.

Pellets (*Fig. 120*) are a relatively new form of wood fuel. The pellet is the name for the granule of circular cross-section with a diameter of about 6–8 mm and a length of 10–30 mm. They are made exclusively from waste material such as sawdust or wood shavings without any chemical additives. The density achieved by high pressure is similar to that of wood briquettes. The pellets are characterised by the moisture content of about 8–10 % and the heating value of 17–18 MJ.kg^{-1}. With these parameters, pellets compete with low-coalified coal. Wood pellets most often consist of three main building elements – tracheids (*Fig. 120b*), parenchymatic and sclerenchymatic cells – based on the origin of wood, from which they are produced.

a)   *b)*

Fig. 120 Wood pellets (a) and tracheids in their structure (b)

11.1.2.2. Phytomass

Herbaceous biomass (phytomass) is produced by plants that have a non-wood body. Similarly to wood biomass, phytomass may also be used directly in the raw or modified form (pellets, briquettes and pressed packages). However, phytomass as energy fuel is also characterised by diverse drawbacks. Although the calorific value of phytomass is only slightly lower than that of wood, the ash content of such fuel is about four to ten times higher. Cereals and grains also have a much higher content of nitrogen, potassium and chlorine in comparison with wood (as a result of fertilising and spraying). The mentioned elements are involved not only in the production of higher emissions but to a large extent influence the corrosion in combustion plants, as well as the creation of clinker due to lower softening and melting temperatures. Most agricultural biofuels have a low point of softening, melting and creep of ash, and they also have a large production of CO.

There are considerable differences among individual types of phytomass in their chemical composition and structure of matter, e.g. chemical composition of rape straw and its seed is very similar. Nevertheless, there are some differences affecting their quality. It concerns particularly sulfur content, which amounts to 0.6 % in the seed-residue. It is a figure corresponding to values found in the coke powder, *Tab. 34*. The ash after combustion of sugar cane pressings has strongly acidic character, and SiO_2 constitutes more than half of its weight, *Tab. 35*.

The structure of sunflower husks is not homogeneous and is characterised by the anisotropy given by its growth characteristics. According to *Fig. 121*, the system consists of small, longitudinally arrayed hollow fibres [87, 88]. They multiply the reaction surface after grinding due to their shape. An interesting feature of chemical analysis regarding sunflower husks is a high concentration of alkaline oxide K_2O in the ash reaching more than 30 %, *Tab. 35*. This component is predominant and approximately equal to the sum of basic oxides of CaO + MgO.

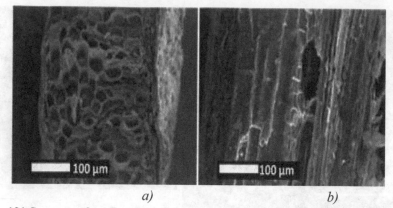

a)   *b)*

Fig. 121 Structure of sunflower husk in cross (a) and longitudinal section (b) [87, 88]

Thermally untreated phytomass is characterised by soft and fibrous nature, and it cannot be ground by methods used for grinding of coal or coke. The milling apparatus should be based on the shearing action, where the material is crumbled by the contact of rotating blades and a fixed rail.

Straw crops, mainly cereals and rape, are the most significant source of biomass in agriculture, which is used for energy purposes. In terms of burning straw, the most important feature is its quick and easy gasification. At temperatures of about 200–250 °C, the major part of the material is gasified. The structure of the straw is causing the escape of fine ash. It is made up of both unburned or charred tiny particles of straw and ash, which tend to be partially bonded to the heat exchange surfaces and partly escape to the atmosphere. It is therefore very important that the straw is not in contact with the refractory lining during the entire heat treatment process. The ash, which consists of minerals with low melting point, begins to soften at about 830 °C. At temperatures of 850–900 °C, it easily forms a glassy substance that damages the lining and is difficult to remove.

Fig. 122, 123 show the results of derivatographic analysis of four types of phytomass (sunflower, maize, wheat and rape) at different particle sizes [44]. Derivatographic analysis was performed under inert conditions. During heating, gaseous components (mainly volatile combustible) are released in the following temperature ranges for different types of phytomass:

- 260–480 °C (sunflower),
- 260–500 °C (corn),
- 320–620 °C (wheat),
- 330–750 °C (rape).

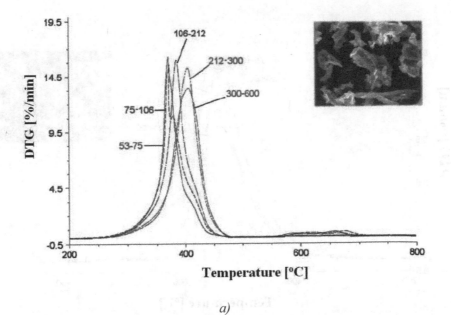

a)

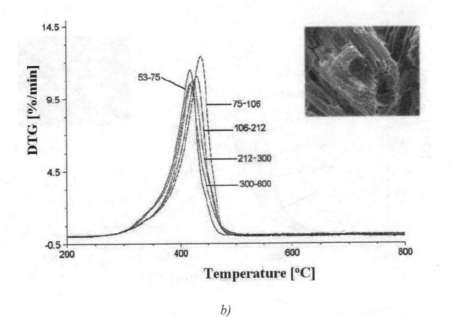

b)

Fig. 122 Data from derivatographic analysis of individual types of biomass with different particle sizes (in μm)
[44]
a) sunflower, b) corn

From the DTG results, it is clear that the rate of weight loss in the sunflower, corn and wheat is the highest in the lowest grain size classes (53–75 μm and 75–106 μm). In the case of rapeseed, the rate of weight loss is the highest for the highest grain size classes (600–1180 μm and 300–600 μm).

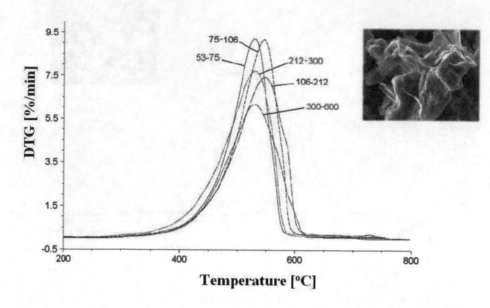

a)

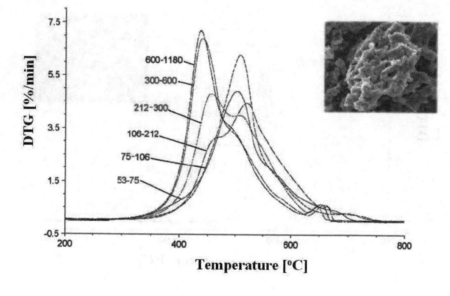

b)

Fig. 123 Data from derivatographic analysis of individual types of biomass with different particle sizes (in μm)
[44]
a) wheat, b) rape

Hemp can be also included in the plant biomass. The substance of the hemp stalk mass is formed by longitudinal fibres of different sizes arranged closely side by side, *Fig. 124*. A sample of hemp pulp (stalks) can be visually observed without preparation.

a)   *b)*

Fig. 124 Picture of hemp sample (a) and optical observation of hemp sample (b)

Conglomerates of hemp material have the appearance of short rods with a diameter of 20 to 50 μm and a length of 100 to 300 μm. Chemical composition in EDX analysis differs from the chemical composition of, e.g., wood biomass in each analysis point. Most of the particles are characterised by the dominance of calcium and potassium but some particles are characterised by a high content of phosphorus and magnesium, *Fig. 125*.

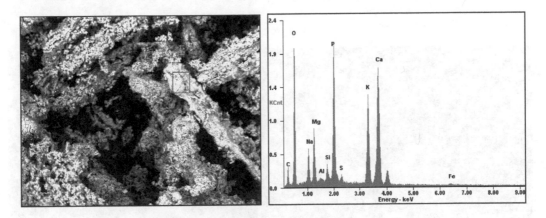

Fig. 125 EDX analysis of hemp

11.1.2.3. Fruit biomass

Biomass created by plants in the form of fruits is assigned to a specific group of plant biomass. These raw materials are almost exclusively processed by the food industry, and residues from the production are available for use in energy recovery. The use of hazelnut shells, almond shells and walnut shells is very promising. *Fig. 126a* shows a particle of a hazelnut shell, which forms an aggregate composed of many homogeneous, slightly elongated, and at the ends rounded parts [88, 89]. There is no discernible porosity but there are microgaps among individual parts. Hence, the surface properties remain more or less unchanged even with finer grinding and enlargement of the reaction surface. Almond shell is made up of bodies without the similarity of shape, and microporosity is not discernible, *Fig. 126b*.

a)   b)

Fig. 126 Structure of certain types of fruit biomass [88. 89]
a) hazelnut shell, b) almond shell

Microstructural analysis of walnut shell has shown that a substantial part is formed by the parenchymatic cells with the lowest porosity, *Fig. 127b*. Samples of walnut and almond shells were observed without modifying their particle size composition – i.e. in their original form.

a)   b)

Fig. 127 Sample of walnut shells (a) and their optical observation (b)

The apparent size of the walnut shell conglomerates is greater than that of the wood biomass and is in the range of 100–500 μm. The particle size dispersion is significant. Most of the particles contain mainly calcium and potassium, *Fig. 128*.

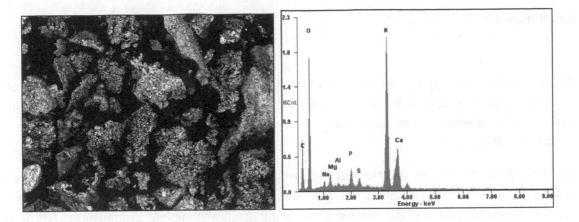

Fig. 128 EDX analysis of walnut shell

11.1.3. Biomass as a source of energy

The use of biomass as a renewable source is of great importance in terms of energy. Its undisputable advantages are particularly the high energy potential and centralised energy production. The possibility of biomass production in local conditions reduces the energy dependence on imported energy sources. A significant advantage is the treatment and disposal of waste, which may be both unprocessed and toxic. There is no doubt about the advantages of waste biomass processing. By such processing, it is possible to dispose of waste that would otherwise not be utilised. It is clear that, at present, biomass may be used in many industrial and energy industries. The question of effectiveness and efficiency of the use of specific biomass in metallurgy must be associated with thorough research of this biomass for the purpose of its comprehensive assessment. The world energy market and energy security are very unstable, and therefore biomass research needs to continue in the future.

For biomass processing, a whole range of chemical processes is employed, changing it into another type of fuel (wood – charcoal) or directly into thermal energy (wood – burning). The thermochemical transformation of biomass in so-called dry processes is used the most:

- combustion,

- gasification,

- pyrolysis.

Combustion

Almost any biomaterial can be used as a fuel for combustion processes. The most common biomaterial is wood and waste from the wood industry, straw, cereals, vegetable waste from agricultural production, organic and municipal waste. Oily crops such as rape or sunflower can also be used.

In combustion, exothermic reactions of fuel with oxygen occur. In addition to carbon dioxide and water vapour, other pollutants also arise during the combustion of biomass. They include solid particulates and carbon monoxide resulting from incomplete combustion. Nitrogen oxides are formed as well – not only from air nitrogen but, above all, from nitrogen present in biomass. Sulfur is also present in the biofuel, which contributes to the formation of sulfur oxides during the combustion process. However, their concentration is significantly lower than that of coal or coke combustion. In the ash, a relatively higher proportion of heavy metals can be found.

In the combustion of farm phytomass, higher chlorine emissions are also produced, which have an aggressive impact on the combustion chamber. In the combustion process, after fuel heating and moisture removal, a mixture of hydrocarbons, carbon monoxide, hydrogen and nitrogenous substances is released, which burns together with fixed carbon. Exothermic reactions in burning (combustion) of these substances release heat energy. The temperatures at which the chemical reactions take place range from 300 °C to 1200 °C. For biomass, a high proportion of volatile fuel is typical. This very proportion defines the requirements for burning time, flame length, the proportion of air oxygen, so that biomass being combusted is burned efficiently. These requirements may cause significant restrictions on the use of specific biomass in the combustion process. The energy use of biomass has a negligible effect on the production of so-called "greenhouse gases" because it is assumed that the amount of CO_2 emitted by the burning of a certain amount of biomass has already been accumulated. For this reason, the emission factor for biomass burning is often incorrectly stated as zero. Although it is the correct value from the point of view of the European Union legislation for the calculation of CO_2 emissions, from the physical point of view, CO_2 emission factors range from 62 to 86 kg/TJ.

Gasification and pyrolysis

For gasification or pyrolysis, mainly waste from woodworking industry and organic municipal waste are used. During gasification, a partial chemical conversion of the solid fuel into gas is accomplished by means of gasifying substances (water vapour, O_2, H_2 and CO_2). The temperatures at which the chemical reactions take place are about 700 °C to 1100 °C. The gas which is the product of gasification contains H_2, CO, CH_4 and other hydrocarbons. It can also contain nitrogen (either airborne or from biofuel), oxygen, as well as a small amount of pollutants such as tar, dust, oxides of nitrogen and sulfur.

The biomass gasification is the reaction of carbon with water vapour, oxygen or both of these gaseous media.

The endothermic reaction is the basis of gasification:

$$C_{(s)} + H_2O_{(g)} = CO_{(g)} + H_{2\,(g)} \qquad \Delta H_{298} = 131.285 \text{ kJ} \qquad (35)$$

The biomass gasification by a mixture of water vapour and oxygen is more preferable, where two exothermic reactions (reactions 7–8) are added to the endothermic reaction.

Other important chemical reactions that occur during gasification are the formation of methane (36) and Boudouard reaction (9).

$$C_{(s)} + 2\,H_{2\,(g)} = CH_{4\,(g)} \qquad \Delta H_{298} = -74.600 \text{ kJ} \qquad (36)$$

The formed gas may also contain methane. Its content is increasing with the decreasing temperature and the increasing pressure (usual temperature at gasification of the biomass is 800 °C, and usual pressure is 2–3 MPa). A well-known biomass gasification plant is located in Güssing, Austria. It gasifies pure wood chips from the surrounding forests, and the produced gas is combusted in internal combustion engines. The heat is supplied to the remote heating network and is used for drying furnaces. The electricity is supplied to the distribution grid.

Pyrolysis is a simple and probably the oldest way of processing biomass for higher quality fuel (charcoal). Pyrolysis is the thermal decomposition of the mass without air. Temperatures range from 400 °C to 700 °C, producing gas that is a mixture of hydrocarbons, H_2 and CO. Solid component with high carbon content (wood charcoal) is another major product of this process. Hemicellulose, cellulose and lignin, which are the main structural components of wood, cause acceleration of pyrolysis in the temperature range of 180–550 °C. The great advantage of pyrolysis is that it takes place without oxygen and does not generate dioxins, which are hazardous substances generated by the combustion of waste containing chlorine.

At present, pyrolysis is considered a very promising technology. It is related to the fact that the process takes place at relatively low temperatures, leading to lower emissions of potential pollutants compared to the conventional biomass burning. Pyrolysis is used for the environmentally sound disposal of waste and has an added value in providing usable products: pyrolysis coke, pure carbon, charcoal, bio-oils, pyrolysis gas, etc.

11.1.4. Biomass and the environment

Solid pollutants, sulfur oxides, oxides of nitrogen, carbon monoxide and greenhouse gases (carbon dioxide, water vapour and methane) belong to the most significant emissions emitted into the air by biofuel combustion. Other pollutants such as heavy metals, hydrogen fluoride, halogen compounds and unburned hydrocarbons are emitted by the combustion of these carbonaceous fuels into the atmosphere in smaller quantities [90]. Due to the toxicity and persistence, these pollutants have a very negative effect on vegetation, animals and humans.

Solid (dust) particles in the air are a mixture of substances consisting of carbon, dust and aerosols. A number of terms are used for their identification. We can come across the terms solid polluting matter, solid aerosol and airborne dust. In the foreign literature, the terms suspended particulate matter (SPM), total suspended particles (TSP) and black smoke are used.

The particle size and composition of the particles are related to the effects of the particles on human health and possible risks to the exposed population. At present, the greatest attention is paid to particulates with an aerodynamic diameter below 10 μm (PM10) that can penetrate into the respiratory tract. The biggest threat is the smallest 2.5 μm (PM2.5) dust particles, which are able to last for several days in the stratospheric layers and even for several years in the stratosphere, allowing them to travel over long distances. Airborne dust particles (aerosols) often act as condensation cores and have a demonstratable effect on cloud formation and precipitation, and thus also affect weather and climate. Particulates having an aerodynamic diameter below 10 μm and shares up to 2.5 μm are typical examples of solid pollutants originating in biomass heat treatment processes.

A more detailed analysis of the relationship between the alternative fuels (including biomass) and the environment is described in chapter 15 "Alternative Fuels and the Environment".

11.1.5. Methods of using biomass for metallurgical purposes

For the production of pig iron in a blast furnace (BF), blast furnace coke is technologically replaceable, but only to a certain extent. Biomass can replace a part of coke in the BF in terms of its calorific value and as the reducing agent. By gasification of biomass, it is possible to obtain hydrocarbon gas fuels, which can be injected through tuyeres into the blast furnaces.

Use of biomass within the framework (for stability) of the blast furnace charge is still unrealistic.

In the agglomeration process, use of different types of biomass as a substitute for coke powder is prospective. The properties of coke powder are closely related to pyrolytically processed biomass – charcoal. From the chemical composition point of view, the advantage of wood charcoal is the lower ash and sulfur content. Carbon content is comparable to coke powder. The negative factor is the higher content of water and volatile substances. Increased biomass moisture is a typical feature, which significantly affects the combustion process and is mostly negative. It reduces the calorific value of the fuel, as a part of the energy is consumed to evaporate the water contained in the fuel. It reduces the efficiency of the burning process, resulting in the formation of a larger unburned residue in ash, which is without energy effect. The energy properties of biomass are a decisive parameter, which refers to its suitability for use as an alternative fuel in the agglomeration process. It is evident from comparing the energy values of coke powder and charcoal that their calorific values are at comparable levels. However, the cost of biomass treatment is fully transferred into the fuel price, which will ultimately be a decisive indicator of its applicability in the agglomeration process.

Coke powder shows significantly more pronounced differences in comparison with raw biomass, i.e. oak and pine wood sawdust. Wood biomass is characterised by a higher water content, higher volatile content and lower carbon, which is a negative feature from the energy point of view. The result of these biomass properties is about half of the calorific value compared to the reference coke powder. In the complex evaluation of dendromass, the sulfur content is very favourable, which is approximately 10-times lower than that of the reference coke powder. More significant differences can also be noticed in the quantity and composition of ash. Biomass contains 10 to 15-times less ash than the reference coke powder. At the same time, biomass is the major alkali carrier, $K_2O + Na_2O$. Wood biomass is characterised by acidic ash and contains higher amounts of alkali oxides that are not desired in the agglomeration process. However, the positive factor is that the ash content in biomass is very low compared to coke powder. The importance of thorough and comprehensive biomass research lies in a detailed understanding of its composition and structure as a starting point for finding a renewable energy source.

The production of ferroalloys is based on reduction processes. The efficiency and quality of ferroalloy production in electric arc furnaces is very much dependent on the quality of the reducing agents used for this production. In practice, bituminous coal and coke are used as major carbonaceous reducers for the production of ferroalloys. Wood chips and sawdust are also used in many factories, while their main task is to increase the permeability of the charge. In addition to traditional reducers, some types of biomass are used for reduction, e.g. charcoal. However, the use of charcoal is significantly limited by its higher price in comparison with fossil fuels, and its efficient use is rather localised (e.g. in Brazil).

12. Utilization of biomass in the production of Fe agglomerates

Either partial or full substitution of the agglomeration coke by other, less environmentally demanding and economically viable fuels appears to be a possible solution to reducing energy emission and thus the economic burden of agglomeration process of fine-grained iron-bearing materials. The reduction of the emission load on the environment can be achieved by the suitable quality of charge raw materials, as well as by control of selected sintering parameters (sintered layer height, volume and velocity of sucked air, quantity and type of fuel, the ratio of concentrate to sinter ore, etc.). The replacement of coke powder as agglomeration fuel, although so far only partial, is becoming an issue that has been ever more topical in recent years. The aim is primarily to circumvent the carbonisation process and thus significantly reduce the cost of treatment of fuel raw materials, which ultimately translates directly into the price of agglomerated charge for the blast furnace process. Also, the tightening of the energy emission limits emphasises another characteristic of agglomeration fuel, which traditional fossil types cannot have, namely the environmental acceptability. The most preferred resource in this area is plant biomass, which is neutral in terms of greenhouse gases emissions, and therefore represents an attractive opportunity as a fuel substitute for industrial use.

12.1. Replacement of coke powder with biomass

The ways of using biomass for energy purposes are predominantly predetermined by its physical and chemical properties. Within the available studies from abroad, the following types of biomass were tested in the agglomeration process (their granulometric composition is given in parentheses as well) [91–96]:

• sunflower husks and briquettes (0.5–1 mm),

• sunflower husk pellets (4–2 mm), (2–0.85 mm), (0.85–0.60 mm),

• pellets of pressed sugar cane (2–8 mm).

• corn cobs (< 3 mm),

• bark from fir and spruce (< 3 mm),

• sawdust from oak and beech (< 3 mm),

• pressed sawdust (< 4 mm), (< 2 mm),

• residues of olives,

• shells from almonds (< 2 mm),

• hazelnut shells (< 2 mm),

• straw from rape,

• dry straw,

• rapeseed,

• wood charcoal (< 3 mm), (< 8 mm).

In the Slovak Republic, the following types of biomass [97–111] were tested in the agglomeration process:

• fir and spruce bark (< 3 mm),

• sawdust from oak, beech and pine (< 3 mm),

• corn cobs (< 3 mm),

• walnut shells (< 2 mm),

• wood charcoal (< 3 mm).

The results of the study show that the size of wood and the plant biomass should be less than that of coke powder where the limit is below 3 mm. Charcoal should be used in a fraction below 3 mm. Sawdust should be used in a fraction below 1 mm and the value for plant biomass is 0.6–0.8 mm. Research of charcoal as a substitute fuel for the sintering process was carried out by R. Lovel, et al. [84]. Charcoal made of Australian wood Acacia nilotica, Eucalyptus camaldulensis and Pinus nigra were studied. All material was ground to < 3 mm, and the fraction < 1 mm was removed. T.Z. Ooi, et al. used hardwood charcoal in his study [93]. Granulometry of briquette milling was determined to be < 3 mm, and grains smaller than 0.85 mm were not used.

The biomass is characterised by the relatively high and frequently variable water content, which significantly affects the energy properties of biomass fuel. The various types of biomass (primarily of wood and plant origin) have much higher content of volatile matter than coke powder. Analyses of the various types of biomass that have been used in the production of iron-bearing agglomerate worldwide are shown in *Tab. 37*. It follows from the chemical analyses that some types of biomass have a very high amount of volatile matter (e.g. sawdust, corn, rape, etc.) and constitutive water, which is bonded to a volatile combustible substance. Some materials are easily spoiled (e.g. sawdust) and contain oily particles (e.g. rapeseed), which are not suitable for the agglomeration process. Furthermore, these oily particles may cause problems in an electrostatic precipitator.

Tab. 37 Analysis of selected carbonaceous materials which have been used in agglomeration worldwide [91–96]

Carbonaceous materials	Moisture [%]	Ash [%]	Volatile matter [%]	Fixed carbon [%]	Sulfur [%]	Calorific value [MJ/kg]
Coke powder 1	5.5	12.1	1.5	84.4	0.60	28.00
Coke powder 2*	-	19.5*	5.8*	74.6*	0.50*	26.84*
Sunflower husks	9.3	3.2	76.4	12.3	0.14	16.00
Hazelnut shells	11.0	1.1	67.8	20.0	0.04	18.20
Almond shells	9.3	0.5	81.4	8.8	0.03	16.70
Rape 1	9.8	2.8	81.2	6.0	0.11	17.82
Rape 2	8.1	4.5	83.2	3.6	0.60	17.37
Sawdust*	-	1.3*	84.2*	14.5*	0.02*	17.22*
Charcoal 1	4.6	4.5	32.7	58.2	0.02	33.10
Charcoal 2*	-	5.1*	7.5*	87.3*	0.04*	30.77*

Note: Analysis of ash, volatile combustible, fixed carbon and sulfur was carried out on a dry sample.
    * Dry state

An analysis of individual types of biomass that have been used in the production of iron-bearing agglomerate in the Slovak Republic is shown in *Tab. 38*. Given the nature of origin and theoretical knowledge of cell mass constitution of biomass and coke, significant differences in the chemical composition of compared fuels can be assumed. Fixed carbon content in the tested biofuel combustible (except for charcoal) was about four times lower compared to coke powder. Thanks to the photosynthetic reaction, carbon in biomass is bound primarily to hydrogen and oxygen in the form of cellulose macromolecules ($C_6H_{10}O_5$) with a certain degree of polymerisation, which plays a major role in the noticeable difference in the amount of volatile matter. Volatile matter content in various types of biomass (besides charcoal) is about 75–85 % (in the case of coke powder values, it is about 2–4 %). The content of solid as well as total carbon in charcoal is at the level comparable to coke powder. The content of volatile matter in charcoal is slightly higher than in coke powder. The details of chemical composition given above and their effect on the energy potential have been confirmed by the measured calorific value. It showed the highest calorific value of charcoal (CV = 30–33 MJ/kg) and coke powder (CV = 28–29 MJ/kg). The heat content of all other tested types of biomass was approximately by 40–45 % lower than the heat value of coke powder.

Tab. 38 Analysis of selected carbonaceous materials used in agglomeration in the SR [97–111]

Carbonaceous materials	Moisture [%]	Ash [%]	Volatile matter [%]	Fixed carbon [%]	Sulfur [%]	Calorific value [MJ/kg]
Coke powder 1	1.5	14.5	3.5	82.0	0.59	28.16
Coke powder 2	0.1	13.3	1.8	84.9	0.51	28.75
Walnut shells	9.6	0.7	81.3	18.0	0.05	16.90
Spruce bark	7.3	3.0	75.7	21.3	0.02	18.68
Beech sawdust*	-	0.3*	85.4*	14.3*	0.01*	17.53*
Oak sawdust	7.1	1.5	83.4	15.1	0.05	16.56
Pine sawdust 1	13.6	0.9	85.6	13.5	0.05	15.94
Pine sawdust 2*	-	0.9*	85.2*	13.9*	0.05*	18.93*
Corn*	-	1.6*	83.2*	15.2*	0.05*	16.11*
Charcoal 1	18.0	6.1	17.0	76.9	0.02	29.93
Charcoal 2	4.9	3.5	8.2	88.3	0.05	30.46
Charcoal 3*	-	2.3*	6.4*	91.3*	0.05*	32.66*

Note: Analysis of ash, volatile combustible, fixed carbon and sulfur was carried out on a dry sample.
    * Dry state

The ash content in the tested samples of anhydrous biomass (except for charcoal) was very low (0.7 to 3.0 %) compared to coke powder (13.3 to 14.5 %). It also means a significant difference in sulfur content brought to the process because its content in the anhydrous sample of coke powder (about 0.5 to 0.6 %) is considerably higher than in samples of biofuels (0.01 to 0.05 %). The ash content in charcoal is higher than in the raw biomass but significantly lower than in coke powder. The ash, which is defined as the solid unburned residuum of fuel, is formed mainly by oxides of mineral substances contained in the biomass (SiO_2, Al_2O_3, Fe_2O_3, CaO, MgO, K_2O, Na_2O and P_2O_5), **Tab. 39**.

In general, ash content in biomass fuels is much lower than in other types of solid fuel, thus ensuring the smaller content of particulates in the flue gas. The ash content in biofuels is given by the proper chemical structure of the various types of biomass, as well as by the contamination of soil from which biomass drew nutrients for its growth. The results of ash analysis show that the highest levels of SiO_2 are in sawdust (oak and pine) and corn. The highest contents of CaO are in beech sawdust, in spruce bark and in charcoal 2. Charcoal 2 ($p_2 = 6.9$) and walnut shells ($p_2 = 9.9$) have a marked basic nature. Oxides of alkali metal represent another undesirable component of ash. They pass into the agglomerate, and thus ultimately into the blast furnace process, which contributes to the worsening of the technological process and performance parameters of the furnace. The highest K_2O content in ash samples was analysed in corn and walnut shells.

Tab. 39 Chemical composition of ash from biomass fuels (SR) [83, 99, 104]

Type of fuel	Ash content [%]	Chemical composition of ash* [%]						
		SiO_2	Al_2O_3	Fe_2O_3	CaO	MgO	K_2O	P_2O_5
Coke powder	14.5	34.7	21.1	27.2	6.8	2.8	1.6	0.6
Walnut shells	0.7	2.8	1.0	4.2	33.1	5.2	17.8	3.7
Spruce bark	3.0	34.4	6.9	5.0	39.5	4.8	7.1	-
Beech sawdust	0.3	9.2	5.0	24.6	40.0	8.5	6.6	-
Oak sawdust	1.5	41.1	6.7	4.5	23.8	2.6	9.3	0.72
Pine sawdust	0.9	46.6	11.7	6.5	15.2	3.0	5.6	0.85
Corn	1.6	41.1	1.6	2.1	2.6	7.8	37.6	-
Charcoal 1	6.1	22.9	3.2	31.9	28.8	3.0	8.7	-
Charcoal 2	3.5	6.3	0.85	1.47	37.0	12.5	11.42	1.65

Note: * Calculated to ash content

Typical compounds that are found in carbon fuel ash are mullite ($Al_6Si_2O_{13}$), quartz (SiO_2), hematite (Fe_2O_3), calcite ($CaCO_3$) and hedenbergite ($CaFeSi_2O_6$). **Tab. 40** shows the phase composition of ash of some carbonaceous fuels. The phase composition of fossil fuel and dendromass ash is in accord with its chemical composition, when the majority compounds are based on minerals mullite and quartz in fossil fuel (coke) ash. In the case of dendromass (wood charcoal and sawdust oak) ash, it holds true when the majority compounds are based on minerals calcite, magnesite and calcium oxide. In the ash of pine sawdust, a high proportion of quartz was found. A high proportion of amorphous share in oak sawdust is interesting, which can be related to hardly identifiable minerals in this type of dendromass.

Some types of phytomass in the form of waste products from the food industry and agriculture as a replacement for fossil fuels were analysed by M. Zandi, et al. [91]. It concerns residues from oil extraction, rape hay and seeds (Brassica napus), sugarcane (Saccharum officinarum) and sunflower seed husks (Helianthus annuus). The oil extraction moulds as residues from biofuel production were considered unsuitable for agglomeration due to the high volatile content (83.2 % with the H_2 content of more than 8 %). This material is also degrading easily to fine, oily particles that are not suitable for packing into micropellets. At the same time, oily parts would cause problems in gas cleaning in the electrostatic precipitator.

Tab. 40 Phase composition of ash from carbonaceous fuels (SR) [83, 104]

Identified phase composition		Coke	Charcoal	Oak sawdust	Pine sawdust
Chemical formula	Mineral name	[wt%]	[wt%]	[wt%]	[wt%]
$(Ca_{0.94}Mg_{0.06})CO_3$	Calcite	-	57.3	-	15.7
$MgCO_3$	Magnesite	-	26.8	12.1	-
$Ca_6Mn_6O_{16}$	-	-	15.9	-	-
$Al_{1.25}Si_{0.75}O_{4.87}$	Mullite	50.4	-	-	-
$CaFeSi_2O_6$	Hedenbergite	4.3	-	-	-
$CaSO_4$	Anhydrite	8.0	-	-	-
SiO_2	Quartz	16.9	-	10.2	54.9
$Ca_2Fe_{1.54}Al_{0.46}O_5$	Brownmillerite		-	-	10.2
Fe_2O_3	Hematite	16.5	-	-	10.9
Fe_3O_4	Magnetite	4.9	-	-	-
MgO	Periclase	-	-	-	6.1
$CaCO_3$	Calcite	-	-	56.0	-
CaO	Calcium oxide	-	-	19.8	-
Amorphous share	-	19.6	41.0	84.0	44.0

The study of the use of husks from sunflower seeds in the agglomeration was conducted by T.C. Ooi, et al. [94]. These husks came from the production of cooking oil, and after extraction, they were dispatched to the market pressed into briquettes. Their calorific value was 1.75-times lower than in coke powder. Therefore, 1.75 kg of this material should be used when replacing 1 kg of fuel in a sintering mixture while maintaining the heat content.

M. Zandi, et al. considers the fruit biomass – along with other food by-products available in large quantities in the UK – the most appropriate biomass for agglomeration. The research focused on the shells of hazelnuts (Corylus) and almonds (Prunus dulcis) [91].

The comparison of the properties of different types of biomass used in the agglomeration process with the characteristics of standard coke powder according to the analyses within foreign studies showed the following [79, 80, 84, 85, 91–96]:

- on average, the calorific value of the biomass used (excluding charcoal) in the dry state is about 60–65 % of the calorific value of dry coke,

- the calorific value of charcoal is equal to coke powder (maybe even higher),

- types of biomass have a higher reactivity and lower thermal capacity than coke powder,

- types of biomass have a lower density and higher specific surface area,

- charcoal and certain types of biomass (e.g. dry straw) have a higher porosity than coke, and their pores are primarily micropores.

According to the studies, individual types of biomass have (compared to coke powder):

- higher moisture,

- lower total and fixed carbon content,

- higher content of volatile matter,

- significantly lower ash content,

- lower sulfur content,

- higher phosphorus content,

- higher CaO and MgO content, higher basicity,

- higher alkali content.

12.2. Thermodynamic study of biomass utilisation within the agglomeration process

The production of ferriferous agglomerate also requires an analysis of thermodynamic and kinetic conditions of the process. In the world, the study of the thermodynamic and kinetic conditions of solid biomass combustion is mostly carried out in more detail in other technological processes than the production of ferriferous agglomerate using biomass. The results of these studies can be compared, but for practical use of thermodynamics, only specific model cases are relevant. Gibbs equilibrium diagrams, which characterise the change in the equilibrium composition of reactants and reaction products with a temperature change, are calculated to determine the viability of individual temperature-dependent biomass reactions. It is also possible to determine the stability of individual phases at a different partial pressure of the gaseous components using the Kellog diagram of stability areas in the considered solid biomass combustion systems in the agglomeration charge.

Fig. 129, 130 show the modelled course and thermal effect of the individual reactions occurring during the combustion of coke powder (or charcoal or sawdust) during the agglomeration process.

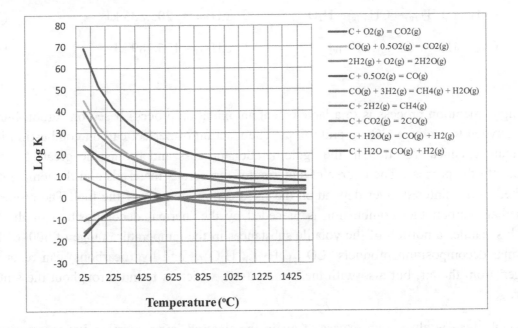

Fig. 129 Temperature influence on course of chemical reactions occurring during combustion of carbonaceous fuels

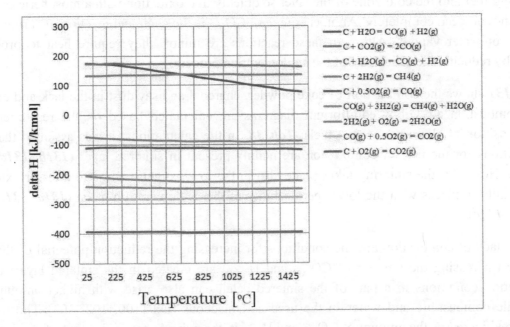

Fig. 130 Temperature influence on thermal effects of chemical reactions occurring during combustion of carbonaceous fuels

In addition to the reactions 7, 8, 9, 10, 35 and 36, the following reactions also take place:

$$2 H_2 \text{ (g)} + O_2 \text{ (g)} = 2 H_2O \text{ (g)} \qquad \Delta H_{298} = -483.652 \text{ kJ} \qquad (37)$$

$$CO \text{ (g)} + 3 H_2 \text{ (g)} = CH_4 \text{ (g)} + H_2O \text{ (g)} \qquad \Delta H_{298} = -205.885 \text{ kJ} \qquad (38)$$

$$C \text{ (s)} + H_2O = CO \text{ (g)} + H_2 \text{ (g)} \qquad \Delta H_{298} = 175.289 \text{ kJ} \qquad (39)$$

The agglomeration process is to a large extent an oxidation process based on carbon-burning with oxygen to produce CO_2 (g) and CO (g). The reaction of perfect carbon combustion is the dominant chemical reaction in the agglomeration process and brings the greatest thermal effect into the process. The course of sintering of ferriferous materials and the temperatures reached in the sintered layer depend on the fixed carbon content of the fuel. The process of immediate carbon fuel combustion is preceded by the thermal decomposition of the fuel, which separates a portion of the volatile substances in the temperature range of 300–650 °C. Thermal decomposition products (CO (g), H_2 (g), H_2O (g) and hydrocarbons) can react with oxygen from the air, but also with the carbon itself, which is found throughout the sintered layer.

Although there is always an excess of air in the sintered layer, burning is never complete, i.e. a certain amount of CO (g) is always present in the exhaust gas. The presence of CO (g) can be explained by the fact that a rapid air flow carries the produced CO (g) on the surface of the burning fuel into the cold zone of the layer so quickly that oxidation with atmospheric oxygen does not proceed completely. Another source of CO (g) is the chemical reaction of carbon with water or water vapour. Although these reactions are minor, they require heat to proceed, thereby reducing the efficiency of the agglomeration process.

Fig. 131 shows the combustion of coke powder, charcoal and sawdust in the lack and excess of combustion air. Gibbs equilibrium diagrams are calculated based on the real chemical composition of all carbonaceous fuels, *Tab. 38*. In the calculations, it was assumed that the conditions for the perfect combustion are mainly present in sintered layer (*131b, 131d and 131f*). However, the sintering takes place also in the low oxygen regime to a lesser extent – especially in places with the lower permeability of the agglomeration batch (*131a, 131c and 131c) 131e*).

In the lack of combustion air, the conditions for increasing the reduction potential of the gas phase (increasing the amount of CO (g) and H_2 (g)) are created in the sintered layer. These reduction conditions in a part of the sintered batch can also arise with higher amounts of volatile combustible and water in the used carbonaceous fuel [100, 102, 103]. The use of charcoal increases the amount of CO (g) and H_2 (g) in the exhaust gas in the case of combustion in the lack of air. It is even more marked when sawdust is used. In the case of combustion in the excess air (standard agglomeration conditions), the use of charcoal decreases the CO_2 (g)/CO (g) ratio.

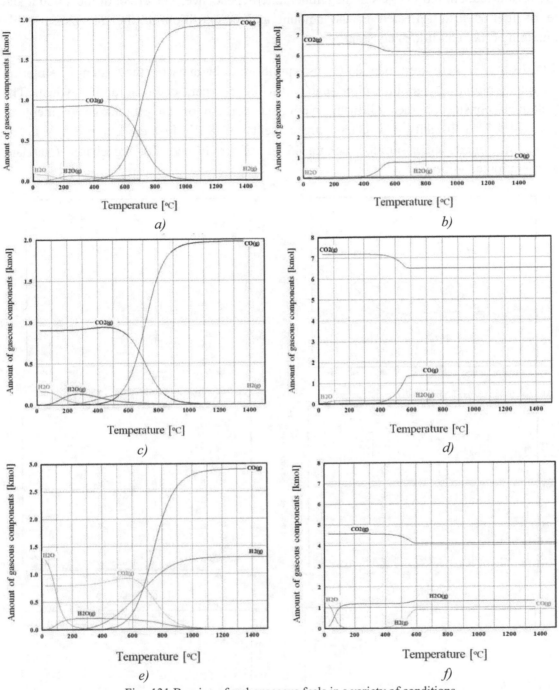

Fig. 131 Burning of carbonaceous fuels in a variety of conditions
a) air/coke powder (1:1), b) air/coke powder (4:1), c) air/charcoal (1:1), d) air/charcoal (4:1),
e) air/sawdust (1:1), f) air/sawdust (4:1)

Gibbs equilibrium diagrams can be used for detailed thermodynamic studies as well. They characterise the temperature dependence of the equilibrium composition of the reactants and products of the considered agglomeration charge sintering system in the presence of coke powder and biomass in various combinations and mixing ratios. Modelled systems and the results of thermodynamic calculations can specify the effect of the amount and type of fuel on oxidation-reduction processes in the temperature dependence. The effect of the amount and type of fuel is also reflected in the amount and composition of the gas phase while maintaining other conditions constant.

In Fig. *132, 133*, there are modelled Gibbs equilibrium diagrams for the system of sintering the agglomeration charge in the presence of coke powder, charcoal and sawdust. The calculations considered the following parameters of agglomeration charge (content of $Fe_{TOT} = 44-45$ %, basicity = 1.6, moisture content = 7-8 %, fuel content = 3.5 %). The sintering was modelled assuming sucking 45 m^3 of air through 100 kg charge (standard agglomeration conditions).

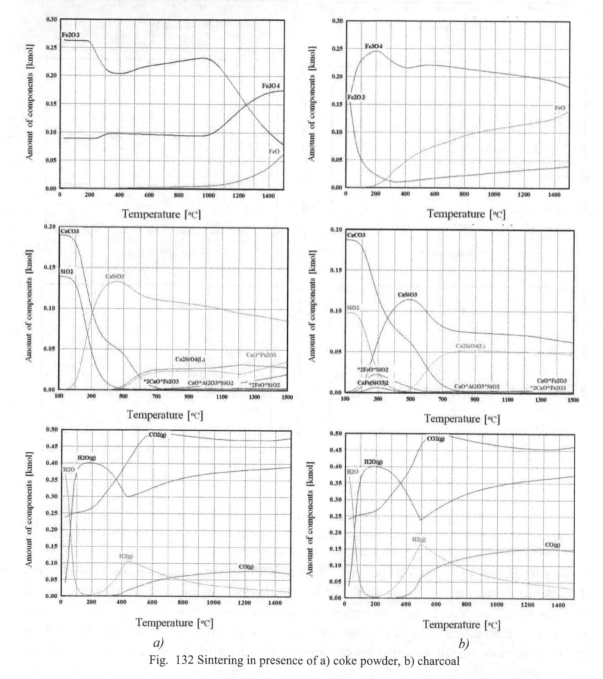

a)   *b)*

Fig. 132 Sintering in presence of a) coke powder, b) charcoal

Based on the analysis of the Gibbs equilibrium diagrams for the system of sintering the agglomeration charge in the presence of coke powder (or charcoal), it can be determined that the substitution of coke powder by charcoal will change the concentrations of the gas components in the exhaust gas, *Fig. 132*. As a result of the higher content of volatile components in charcoal, the course of CO and CO_2 formation reactions tend to decrease their concentration in the gas phase at the expense of increasing the hydrogen content, i.e., steam. In the sintering using charcoal, an increased amount of CO $_{(g)}$ and H_2 $_{(g)}$ is present in the gas phase, which is reflected in a higher content of Fe_3O_4 and FeO in the resulting sinter. On the other hand, in the production of agglomerates using charcoal, it is possible to expect lower amounts of calcium ferrite (modelled by the compounds of CaO and $Fe_2O_3.2CaO.Fe_2O_3$), which can cause the lower strength of the product. Since the coke powder contains a significantly higher proportion of ash than charcoal, a higher proportion of silicate (modelled by the compounds $CaO.SiO_2$ and $2CaO.SiO_2$) and free SiO_2 may be expected in the final structure of the agglomerate (with the use of coke powder).

When sintering using sawdust, the effect of the amount of fixed carbon, the amount of volatile matter and moisture is also thermodynamically reflected in the final thermic effect and the composition of the gas phase, *Fig. 133*.

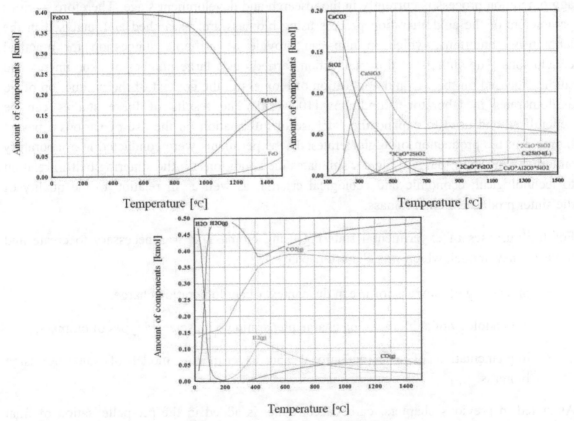

Fig. 133 Sintering in presence of sawdust

In the sintering using sawdust, a higher amount of $H_2O_{(g)}$ is present in the gas phase, and due to the lower carbon content in the sawdust, a lower amount of $CO_{2(g)}$ is present in the exhaust gas. Due to the lower calorific value of sawdust, the temperature in the sintered layer probably will not reach 1200–1300 °C (as in the case of coke powder), which will cause the higher content of Fe_2O_3 in the final agglomerates. On the other hand, using sawdust in the production of the agglomerate may lead to lower amounts of calcium ferrites (modelled by the compounds of CaO and $Fe_2O_3.2CaO.Fe_2O_3$), which can result in the lower strength of the final product.

12.3. Technological aspects of the biomass use

Section 7.1.6 "Modelling sinter production in laboratory conditions and the influence of fuel on agglomeration process" described the generalised knowledge acquired by the authors of this publication in laboratory conditions. It concerned a production of Fe agglomerate using traditional carbonaceous fuels (e.g. coke powder). The issue of the use of biomass within the agglomeration process is currently in the research and development stage. Therefore, various parameters of the agglomeration process using biomass are researched and analysed in the laboratory conditions, which in turn can provide a verifiable suggestion for practical conclusions. The authors of this publication carried out a large number of experiments and studies focused on the use and application of biomass in the context of the production of Fe agglomerates in laboratory conditions [100–111]. The results of these studies can be generalised and used to expand the database of information on the use of various types of biomass in the production of agglomerates. The experiments were conducted in a laboratory sintering pan (LSP) that sufficiently simulates the conditions on the sintering belt in relation to technological, economic and ecological criteria, as well as in relation to the quality of the sinter produced using biomass.

For the purposes of experimental sintering using biomass, it was necessary to create and modify a new model, which was characterised by:

- processing of biomass for use in the context of agglomeration charge,

- methodology of agglomeration charge preparation with selected types of biomass,

- implementation of a computational and experimental model of sintering using biomass.

As stated in previous chapters, carbonaceous fuel is added in the pre-pelletisation of final agglomeration charge to provide the heat source for the agglomeration process. The purpose of pre-pelletisation is the formation of micro-pellets with a uniform distribution of fuel in the agglomeration charge. The formation of micro-pellets is affected by a number of factors, out of which the most important ones are packing properties of the input materials. The study of pelletising properties is based on the conditions necessary for the sintering process, namely the condition of sufficiently pre-pelletised homogenization charge with acceptable permeability. For a partial replacement of coke powder with another alternative biomass fuel, it is, therefore, necessary to take into account pelletising properties of the biomass used. The assessment of the pelletising ability of a material is aimed at evaluating its ability to create bonds with the components forming the pre-pelletised mixture.

The pelletisation cannot be unambiguously defined, as it is influenced by several factors such as particle size distribution, shape and surface properties of grains of pelletised materials, and their chemical-mineralogical composition. The pelletisation should involve the rate of formation and growth of the pellets and their strength properties. The essential characteristic of the assessed pelletising processes is the wettability of the material, which is a force effect between the molecules of a bonding phase and a surface layer of pelletised material. The resultant mutual force acting at the interface of liquid, gaseous and solid phases is characterised by hydrophilicity and hydrophobicity of the material, which is usually quantified by the contact angle. If the limit angle of wetting is in the range of 90–180°, then the solid material is hydrophobic – non-wettable. If the wetting angle is in the range of 0–90°, then the solid material is hydrophilic – wettable. When the wetting angle is 0°, the solid material is characterised as ideally hydrophilic – wettable. It is a force effect of surface tensions arising at the phase interface between the surface of the solid phase, the liquid and the surrounding gas phase. The amount of the surface tension of the liquid will proportionally increase the bond strength between the grains only when the liquid completely wets the surface. The pelletisation is the most often evaluated by two basic experimental tests determining the amount of capillary soaking, using the free drop method. The analysis of pelletising different types and granularity classes of wood sawdust has shown that the highest wettability was achieved by oak sawdust. On the contrary, pine sawdust has the lowest wettability. The most suitable fractions are –0.5 mm (oak sawdust) and 0.5 to 1 mm (pine sawdust). By adjusting the size distribution of biomass, we can optimise its surface properties and, consequently, their pelletising in the context of the agglomeration process.

12.3.1. Processing of biomass

There is little room for extensive pre-treatment of biomass in the agglomeration process. The basic requirement for the use of biomass is to provide the desired particle size composition. It is not possible to carry out treatment (e.g. reduction) of crude biomass in the same way as the treatment of coke powder because biomass has a soft fibrous texture.

Fig. 134 shows the particle size distribution of the tested biomass for the agglomeration process. By the sieve analysis, weight fractions of individual grain size classes were assessed, and cumulative grading curves were calculated and constructed as well. On the basis of the cumulative curves, parameters d_{50}, d_{75} and d_{25} and the uniformity of grain size were determined by the Boltzmann equation of entropy and probability of the state.

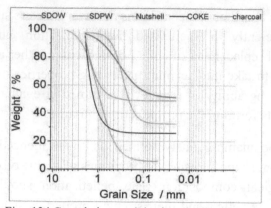

Fig. 134 Cumulative particle size distribution curves
SDOW – oak sawdust, SDPW – pine sawdust

It follows from the analysis of particle size distribution of tested materials that the grain size of d_{50} coke powder was at the level from 1.55 to 1.7 mm. Based on other parameters, it can be specified that 75 % of the total amount has the grain size below 1.9 mm, and 25 % has the grain size below 0.5 mm. The proportion of the coke powder with the particle size distribution below 3 mm was 82–83 %. The uniformity of grain size was determined as the ratio of d_{75}/d_{25}, where the entropy of the sample reached a value of 3.8. The grain size composition of commercial charcoal was adjusted to a particle size close to the coke powder, as the input particle size far exceeded the upper limit of the 3 mm grain size, *Fig. 135a*. Charcoal processing to the desired particle size requires the use of crushing and grinding practices using appropriate calibration pressure system or grid sieves. The time factor (overmilling or undermilling) is no less important to avoid the occurrence of excessive amounts of the defective fraction. After drying the sample of the raw charcoal, the crushing to the fraction under 2 cm followed. The crushed sample is milled in a mill for 30 seconds. The 1.13 to 1.7 mm particle size of charcoal was achieved by milling, corresponding to an average grain size of the tested coke powder. The amount of the proportion under 3 mm in the charcoal was in the range of 75–80 %, *Fig. 134*. After the charcoal grain size adjustment, there was also a tendency to higher levels of fine-grained share compared to the agglomeration coke.

a)   b)

Fig. 135 Photographs of different types of biomass
a) charcoal (untreated), b), charcoal (treated)

Particle size adjustment for the testing in micropelletisation conditions was also performed on walnut shells by the process of crushing and grinding, *Fig. 136*.

a) b)

Fig. 136 Photographs of different types of biomass
a) walnut shells (untreated), b) walnut shells (treated)

After drying the sample of the untreated walnut shells, the crushing to the fraction under 2 cm followed. The crushed sample is milled in a mill for 90 seconds. The particle size distribution adjustment of walnut shells is more difficult. The obtained particle size fractions showed a summary share of 82 % below 3 mm, *Fig. 134*. The share under 1 mm was 21 wt% and the mean particle size was about 2 mm. The granulometric composition of test sample sawdust represented the value of 100 % below 2 mm. The median of the grain size of oak sawdust was 0.56 mm (0.52 mm for pine sawdust). However, oak sawdust showed the lower uniformity of the grain size, manifested by the higher proportion of fine fractions (about 15 % more below 0.315 mm) than in pine sawdust.

The pine sawdust has 93 % share below 1 mm, while the oak sawdust has 85 % of this fraction. When dealing with the particle size of sawdust, it should be borne in mind that the final grain size will depend mainly on the size and shape of the teeth of the cutting tool and its speed, as well as on the composition and structure of the wood.

Hemp straws or hemp pulp were supplied in the modified state in the form of flat, elongated pieces with a length of about 5–20 mm, a width of about 2–5 mm and a thickness of approximately 1–2 mm. The sieve analysis of hemp pulp confirmed these reference dimensions, as the majority share (65 wt%) did not pass through the sieve with a hole diameter of 2 mm, and more than 1/3 of the sample (33 wt%) was collected on the sieve with a hole diameter of 1 mm. Adjusting the particle size of hemp stalks from the delivered state for their further application in the sinter charge in terms of the grain refining poses several problems. Grindability of such fibrous materials is low. It is rather the cutting of the material. Reduction of grain size can also bring other complications in the packed volume of the pre-pelletised mixture, mainly because of its low density and calorific value. Taking into account the coefficient of fuel replacement, a large amount of hemp pulp entering into the pelletising mixture represents a substantial change in the conditions of pelletising (increase in the amount of wetting liquid, increase in the pelletising time, failure to achieve sufficient permeability, reduced bulk density of charge, etc.).

In this section, information related to the adjustment of biomass particle size distribution was specified. But there is also the possibility of modifying the moisture of raw or waste biomass. The most frequently applied system is the adjustment of biomass moisture by drying at about 60–150 °C in a rotary drum dryer. With regard to the biomass, there is low temperature drying at 60 °C, and hot-air drying at about 150 °C. Thus, it is possible to enhance the energy potential of the biomass used. Drying, however, significantly increases the cost of its processing. Therefore, the best solution is to use so-called waste heat for drying. The very moisture of biomass entering the drying process is the parameter that can determine its economic usability.

12.3.2. Methodology for the agglomeration charge preparation using selected types of biomass

To manage the technological process of preparing and processing the agglomeration charge using the biomass, a methodology of sinter charge preparation was developed based on previously learned lessons. This methodology includes technological procedures of biomass preparation and application in the technological flow. These new elements or an extension of technological nodes are necessary to achieve the implementation of a specific substitution in the part of mixture preparation that is optimal from the technological and economic points of view. The incorporation of new technological nodes in a standard process flow required to address the issue of biomass pre-treatment from the technological perspective of mineral processing procedures, regarding the final grain size, capacity requirements for storage, the method of transport, the impact of external factors (climate, length of storage, etc.). The design of the agglomeration charge preparation and processing workflow with the use of biomass is structurally illustrated in *Fig. 137*.

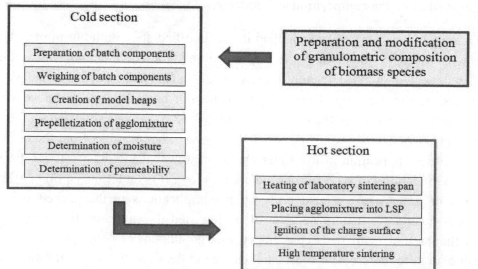

Fig. 137 Scheme of agglomeration charge preparation and processing workflow using biomass

The preparation of charge components includes a possible adjustment of individual commodity grain size, analysis of properties and their storage. With the known performance requirements for the chemical composition of the agglomerate, the individual components of the charge are weighed according to the calculation model of process material balance. The individual amounts of mixed components are stored in the homogenization piles by gradual spreading to ensure the maximum homogeneity of the whole volume of the pile. After the so-called first stage of homogenization, a model pile undergoes the so-called stabilisation for a set number of days. During this time, it is mixed, and a portion of fuel is added as well. This part represents the second stage of homogenization.

The third step of homogenisation is the pre-pelletisation of agglocharge and preparation of final agglomeration charge in blending equipment (e.g. pelletisation drum). In this technological node, another part of the fuel is added before mixing into the charge. The method of fuel addition should ensure even distribution of the addition in the entire volume of the pelletising mixture. The mixture should be in the pelletising equipment in such amount that there are optimum conditions for mixing movement in the drum. The optimal movement of the mixture in the pelletising equipment is important for the formation and growth of nuclei due to the wetting liquid. The addition of substitution fuel requires the determination of optimum pelletising conditions with otherwise unchanged composition of the charge. It is similar to the standard fuel where the optimum particle size and moisture must be determined for a particular charge.

The knowledge of pelletising the mixtures with the addition of biomass shows some differences or effects on technological parameters of pelletising, but also on the process of sintering and quality of the sinter. The optimum conditions for the preparation of sinter charge with the addition of biomass are determined on the basis of experimental results of produced agglomerates and their analysis.

The substitution of coke powder by charcoal generally provides the minimum variation in the preparation of the final agglomixture. The said fuel, after grain size adjustment, is the most similar to the agglomeration coke powder regarding physical-chemical properties. An application of raw biomass (e.g. oak and pine sawdust) in model charges poses greater complications for the pelletisation when the effects of sawdust grain size, as well as wettability of sawdust with water, are noticeable. These parameters have a direct effect on the increase of the pre-pelletisation time, charge permeability, the final moisture content, and bulk mass. These parameters naturally affect the results of sintering conditions and final quantitative and qualitative indicators of sintering.

12.3.3. Execution of computational and experimental sintering model

The creation of a computational model of agglomeration process material balance is based on knowledge of the events taking place in the sintered layer, as well as the principles of conservation of mass and thermodynamic stability of compounds. The model consists of several basic parts. The first part of the calculation model includes analysis of different aggloores, concentrates, secondary ferriferous components, basic ingredients and fuel components. The calculations of the chemical composition of individual components are made from individual components analyses to the so-called delivered state, where the possible moisture is taken into account. The second part of the model is crucial and includes the actual calculation of consumptions of materials in the agglomixture to ensure the desired chemical composition of the agglomerate. The requirement on agglomerate is the compliance with the total content of Fe, FeO, MnO and extended basicity. In addition to the chemical composition of the agglomerate, the results of balance calculations include also mass data of produced agglomerates, which are calculated based on the balance of the chemical composition. The calculations have implemented the above requirements and adjustments from the perspective of various losses and chemical processes during the process. The third part of the model provides a summary overview of the material balance of the process. The computational model is currently being transformed into software and can be verified and also adjusted according to the actual outputs from laboratory sintering.

For the simulation of Fe agglomerate production using biomass, close monitoring of the sintering process needs to be maintained. In section 7.1.6 "Modelling agglomerate production in laboratory conditions and the impact of fuel on agglomeration process", the experimental sintering model has been characterised in detail. The execution of experimental sintering model is also feasible with sintering pan temperature field monitoring by an infrared camera, *Fig. 138*, *Fig. 139* [112]. To measure the temperature of the sintered layer, thermocouples placed along the height of the sintered layer in three measuring zones were used (Zone 1 – 100 mm, Zone 2 – 200 mm, Zone 3 – 300 mm), *Fig. 138* (also *Fig. 75*).

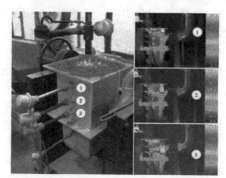

Fig. 138 Infrared image of laboratory sintering pan in individual zones [112]
1st zone) 100 mm – time of 1 minute, 2nd zone) 200 mm – time of 12 minutes,
3rd zone) 300 mm – time of 20 minutes.

a)   *b)*

Fig. 139 Infrared image of ignited charge surface [112]
a) 30 seconds b) 90 seconds

This brings a more detailed picture of the course of fuel burning and thermal wave propagation along the height of the sintered layer, but also in the context of surface ignition of the top layer with an external heat source.

Similarly to the computational model, the experimental sintering equipment has a potential for its improvement and innovation. Most often, it is the automation of the entire sintering process (e.g. automatic charge preparation, automation of ignition head control, automation and software control of negative pressure, etc.). The model of laboratory sintering pan has been further innovated by a transparent high-temperature wall that allows visual monitoring of the combustion zone in the sintered layer during the production of iron ore agglomerate using either traditional or alternative carbonaceous fuels *Fig. 140*.

Fig. 140 Visualization of sintering zone through heat-resistant transparent wall (Slovakia)

The design and implementation of experimental sintering in the laboratory sintering pan was aimed at verifying the thermodynamic assumptions of the behaviour of the biomass types used in the conditions of the sintered layer and the impact of their addition on the technological and qualitative indicators of the agglomerate production. The results of laboratory experiments yielded valuable insight into the impact of the addition of specific biomass types on the agglomerate properties and the emission profile of the gas phase composition.

12.3.4. Impact of biomass on the agglomeration process

Starting from an analysis of the considered biofuels (*Tab. 38*) and proposed methodology for the performance of experiments, model piles were prepared, where agglomeration coke was partly replaced with a defined quantity of individual biomass types. The substitution of coke powder by charcoal was chosen as the first substitute alternative due to the closest similarity of these fuels with respect to the physicochemical properties.

Granulometrically prepared charcoal was applied to agglomixtures and did not have a negative impact on the micropelletisation process, nor on the permeability of the agglocharges, *Fig. 141*. A similar knowledge was also obtained when applying nutshells. The 20 % replacement of coke powder with charcoal and nutshells even increased the permeability of agglocharge. In terms of the agglomeration charge preparation, the addition of sawdust had a negative effect on the formation of micropellets, which was manifested by an increase in the total moisture content and a decrease in the permeability of the agglomixture (*Fig. 141*). In the case of 20 % coke powder substitution by pine wood sawdust, the moisture content of the charge was the highest (about 11 %). It was 4 % increase in the total moisture compared to the moisture of the agglomixture with only coke powder (about 7 %). In the case of 20 % coke powder substitution by pine sawdust, the lowest permeability of the agglocharge was achieved as well.

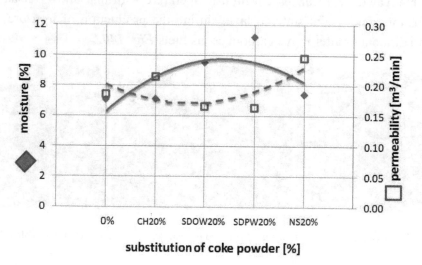

Fig. 141 Change in permeability and moisture of charge in case of coke replacement with biomass
CH – charcoal, SDOW – oak sawdust, SDPW – pine sawdust, NS – walnut shells

The addition of sawdust and nutshells resulted in a drop in the bulk density of the charge, which was mainly affected by the volume of biomass as a result of the substitution coefficient, *Fig. 142*. For comparison, charges with the same percentage of coke substitution have been selected also in relation to the shown impact on charge breathability and moisture, *Fig. 143*.

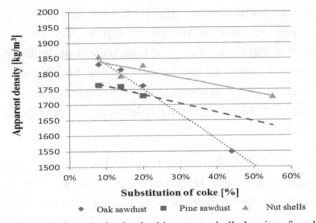

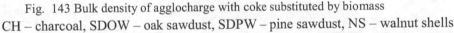

Fig. 142 Effect of coke substitution by biomass on bulk density of agglocharge

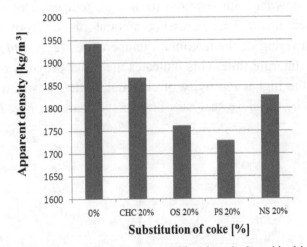

Fig. 143 Bulk density of agglocharge with coke substituted by biomass
CH – charcoal, SDOW – oak sawdust, SDPW – pine sawdust, NS – walnut shells

An increase in water content in the agglomeration charge requires an increase in heat consumption for its transition into the gaseous state. Reducing the permeability of the charge results in an increase in the gas-dynamic resistance of the charge relative to the passage of the gas phase through the column of the sintered layer, which has a direct effect on the course of reactions taking place at individual horizons. The increased gas-dynamic resistance of the charge reduces the intensity of burning, resulting in a lower thermal effect of combustion in elementary layers. Also, the effect of charge permeability is reflected in broadening of burning zones and disharmony of the burning zone with the thermal wave. This can lead to the extreme where the burning zone is interrupted due to the lack of heat necessary to reach the ignition temperature of the fuel. Another extreme may be the non-utilisation (failure to deliver) of the heat generated by the combustion fuel in the charge due to the high velocity of the flowing gas phase as well as the fuel reactivity. Therefore, it is very important to ensure the sufficient permeability of the charge, while its optimum always depends on the particular composition and granulometry of the mixture. In such case, the fuel reactivity plays an important role.

Besides an increase of the water content and reduction of the permeability of the charge, application of both types of sawdust in the agglomeration charge resulted also in the difference in degradation of fuel mass due to differences in the volatile combustible content, which is mostly released at lower temperatures than the burning of fixed carbon in coke. Volatile combustible content (V_{daf}) in sawdust and nutshells is above 80 %, which represents a major difference in the composition and fuel construction compared to coke powder (3.5 %) and charcoal (8.2 %), *Tab. 38*.

Substitution of coke powder by the individual types of biomass thus proceeded with a gradual increase of the substitution coefficient. In addition to the course of the packing, the conditions of the sintered layer regarding the thermal field and changes in the flue gas profile throughout the sintering period were monitored as well.

When replacing coke powder with charcoal (0–86 %), temperatures of 900–1380 °C were reached in the sintered layer. The increasing amount of coke powder substitution with charcoal resulted in lowering of the maximum temperatures (*Fig. 144*) in the sintered layer and shortening of the sintering time. This indicates the faster progress of burning zone in the sintered layer due to the higher reactivity of charcoal, which was also supported by higher dust content, as shown by the sieve analysis of processed charcoal, *Fig. 134*.

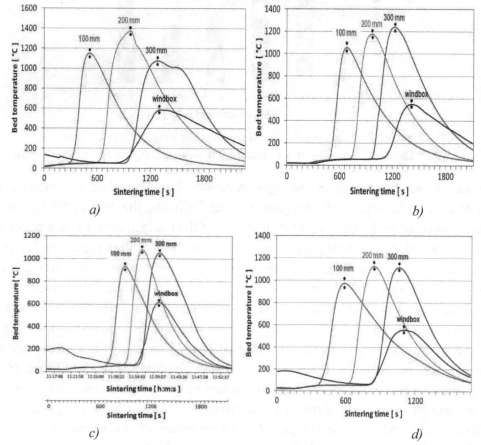

Fig. 144 Temperature curves in measured horizons of sintered layer with % of coke substitution by charcoal (CH) a) 0 % CH, b) 20 % CH, c) 50 % CH, d) 86 % CH

The influence of the rising share of charcoal on the decrease of maximum temperatures was significant in the direction of the sucked air – especially at the lower levels of the sintered layer (200 or 300 mm). The maximum flue gas temperatures for the coke-charcoal substitution (0–86 %) were at 530–620 °C. With the increasing substitution of the coke powder by charcoal, the vertical sintering rate also increased, resulting in the said shortening of the sintering time and subsequent increase in the production, *Fig. 145*.

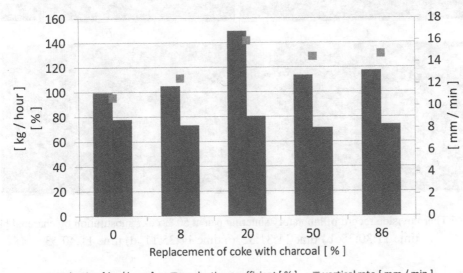

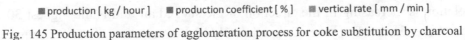

Fig. 145 Production parameters of agglomeration process for coke substitution by charcoal

Fig. 146 shows the thermovision temperature measurement in the laboratory sintering pan at 50 % substitution of coke powder by charcoal. A thermovision camera was placed 5 m from the sintering pan, and the temperature profile was measured at 1-minute intervals. With the thermovision profile of the agglomeration pan, it is possible to monitor the shift of the fuel burning zone in the agglomeration layer. In about 10 minutes, there is a gradual shift of the burning zone from the highest level of the sintered layer (1st zone – 100 mm) to the lowest level of the sintered layer (3rd zone – 300 mm).

Fig. 147 shows the relationship between the temperatures of the steel surface of the pan and the temperatures in the sintered layer. It is clear that there is a relatively high correlation between the temperatures measured in the sintered layer by means of thermocouples (*Fig. 144c*) and the temperatures measured on the surface of the pan by the thermovision camera.

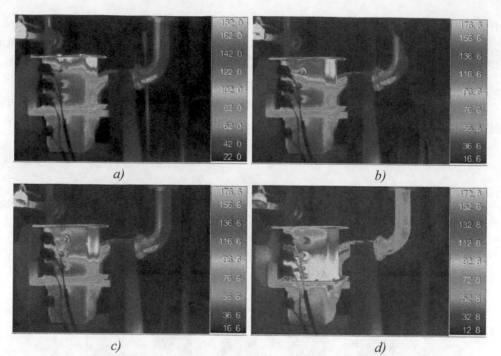

Fig. 146 Thermovision record of laboratory sintering pan at 50 % coke substitution by charcoal [112]
a) time 11:30:38, b) time 11:31:32, c) time 11:35:11, d) time 11:40:53

This correlation is increasing in the direction of the sucked air and is the highest 30 cm from the surface of the charge. Because of the relatively high correlation of the measured values, it is possible to use the thermovision to control the sintering process and to predict the maximum temperatures in the sintered layer.

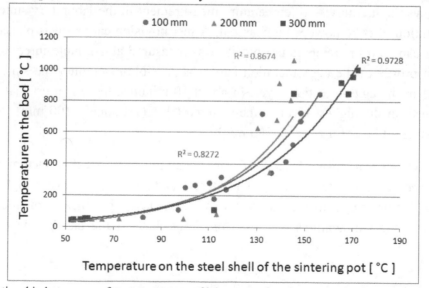

Fig. 147 Relationship between surface temperature of laboratory sintering pan and temperature in sintered layer [112]

With the thermovision camera, temperatures on the surface of the ignited charge and homogeneity of the heat field can be detected as well. The whole ferriferous charge sintering process and the agglomerate quality depends on the uniform ignition of the surface of the agglomeration charge and the sufficient amount of heat supplied by the burner. ***Fig. 148*** shows the thermovision measurement of the temperature profile on the surface of the agglomeration charge in the laboratory sintering pan. The thermovision camera was placed 2 m from the surface of the charge. Temperatures on the surface of the ignited charge were initially in the interval of about 380–540 °C (the ignition temperature of charcoal about 485 °C – is within in this temperature interval as well). These temperatures are lower than the ignition temperature of coke. However, in the case of thermovision measurements, it is necessary to consider the fact that the camera senses an area that is formed not only by particles of burning coke but also by particles of charcoal, iron ore, concentrate and basic additives. Due to the heterogeneous composition of the agglomeration charge, measured temperatures are below the coke burning temperature.

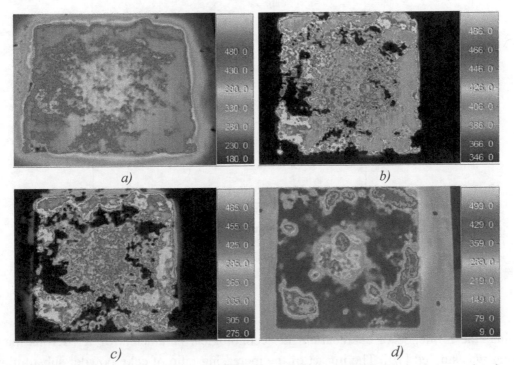

Fig. 148 Thermovision monitoring of laboratory sintering pan surface at 50 % coke replacement by charcoal
[112]
a) time 11:13:55,
b) time 11:14:50,
c) time 11:15:44,
d) time 11:16:52

The measurement results can help to control the charge ignition process and detect its inhomogeneity in the future. ***Fig. 148*** shows that the sintered zone shifted to lower levels in the direction of the sucked air in about 3 minutes. The non-homogeneity of the fuel distribution in the sintered layer and its burning are also highlighted.

In the context of further research worldwide, *Fig. 149* shows the comparison of the temperature profile in the sintered layer using coke powder and charcoal [93]. It is clear that with more than 20 % of the coke powder substitution by charcoal, the maximum temperatures in the sintered layer are lower. The highest temperature of the bed (approx. 1320 °C) was reached in the test with a mixture with 20 % of the coke powder energy substituted by charcoal and subsequent use of 100 % coke powder. The full replacement of coke powder with charcoal has brought the lowest maximum temperatures out of all (approx. 1120 °C). In addition to lower temperatures, the individual temperature profiles also differed considerably. In the case of sintering with 100 % charcoal, a wide temperature profile in the sintered layer was observed compared to other fuel compositions.

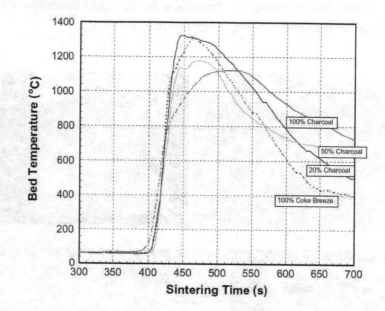

Fig. 149 Temperatures in sintered layer at 0 %, 20 %, 50 % and 100 % replacement of coke powder with charcoal [93]

In China, the influence of charcoal on the technological parameters of ferriferous material sintering was studied [79]. The impact of the increasing ratio of coke powder substitution on the production and qualitative indicators of the agglomeration process is shown in *Fig. 150*. The rising share of charcoal increased the vertical sintering rate, while the yield and the strength index have decreased. With the increase of the charcoal share in the agglomeration charge, the maximum temperatures and the residence times at the maximum temperature are reduced.

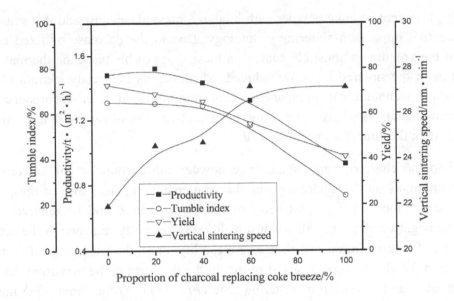

Fig. 150 Effect of coke powder replacement with charcoal on sintering parameters [79]

The effect of other types of biomass on the sintering parameters is shown in **Tab. 41**. These results indicate that the substitution of coke by biomass increases the vertical sintering rate, and on the contrary, reduces the yield, strength index and productivity.

Tab. 41 Influence of biomass types on sintering parameters [79]

Type of fuel	Substitution ratio [%]	Sintering speed [mm/min]	Yield (+5 mm) [%]	Strength index [%]	Productivity [t/(m².h]
Coke powder	0	21.94	72.66	65.00	1.48
Charcoal	20	24.58	68.69	64.40	1.52
	40	24.73	65.30	63.27	1.43
	60	27.20	55.35	54.67	1.32
Dry straw	20	24.05	66.12	63.52	1.42
	30	24.67	63.19	61.33	1.32
	40	25.21	59.56	57.12	1.21
Pressed sawdust	15	23.17	66.21	63.30	1.37
	20	24.56	62.16	61.75	1.30
	40	26.12	54.38	50.11	1.07

At 40 % replacement of coke powder with biomass, pressed sawdust and dry straw have the highest negative impact on sintering technology. Due to the decrease of fixed carbon and increase of the volatile combustible content in these types of biomass, the thermal effect and temperatures in the sintered layer are reduced, which results in already mentioned values of the individual technological parameters. This finding is also in accordance with the thermodynamic analysis where the reaction of perfect burning of carbon delivers the greatest thermal effect in the sintered layer, *Fig. 130*.

Compared to the charcoal application, coke powder substitution by oak sawdust imposes higher requirements on the agglomeration charge preparation. The cause of pre-pelletisation process deterioration is the granulometry of the biomass used and its relatively low bulk density. The negative effect of oak sawdust addition is generally reflected in the mixing time, consumption of wetting liquid and charge permeability. From the technological point of view, the increase in the share of oak sawdust results in the decrease in the maximum temperatures in sintered layer and extension of sintering time (*Fig. 151*). At the same time, the width of the temperature horizons in the sintered layer increases.

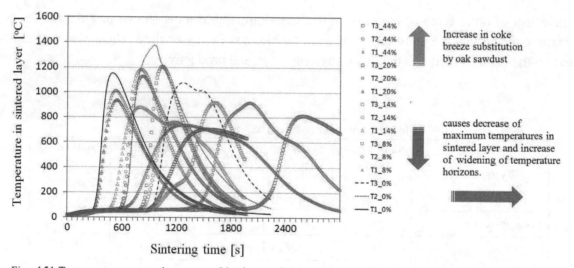

Fig. 151 Temperature curves in measured horizons of sintered layer with coke substitution by 0–44 % of oak sawdust

Fig. 151 shows the differences in the maximum achieved temperatures in the measured horizons of the layer, as well as the changes in the thermal field at individual horizons (the areas outlined by the temperature curves). It is clear from the comparison of the individual curves that with the increasing amount of sawdust the maximum temperature in the sintered layer as well as the speed of the temperature gradient change decreases. It has a negative impact on sintering time, which is extended in the higher substitution of coke powder by oak sawdust. Based on a more detailed comparison of peak temperatures during sintering with pine and oak sawdust, pine sawdust appears to be the worse fuel, *Fig. 152*.

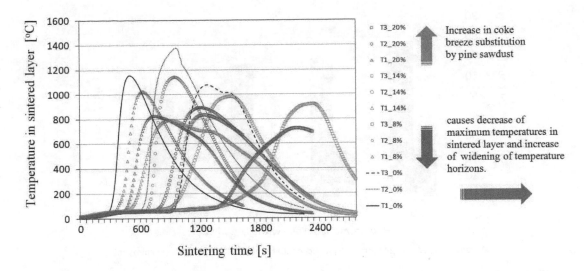

Fig. 152 Temperature curves in measured horizons of sintered layer with coke substitution by 0–20 % of pine sawdust

The comparison of performance parameters when oak and pine sawdust is used in relation to the above impact and temperature profiles indicates the link to the technological and production parameters of the agglomeration process. It is mainly the impact on specific production, the production coefficient, and vertical speed of sintering, **Fig. 153**.

The same trend of the decreasing production factor with the increasing coke substitution is notable for both types of sawdust, while the decrease in the total production with the increasing coke substitution was more marked in sintering with pine sawdust, **Fig. 154**. The values of the calculated vertical velocity, with regard to the functional dependence of production, the factor of productivity and bulk density, changed more significantly in sintering with pine wood sawdust, which also corresponds to the extended sintering time, which has been identified from the temperature profiles and the composition of the gas phase.

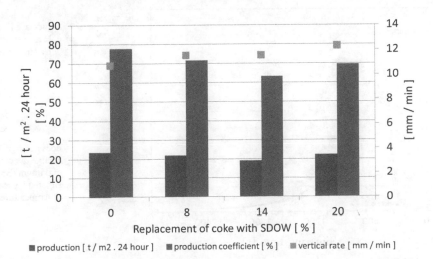

Fig. 153 Production parameters of agglomeration process for coke replacement with oak sawdust

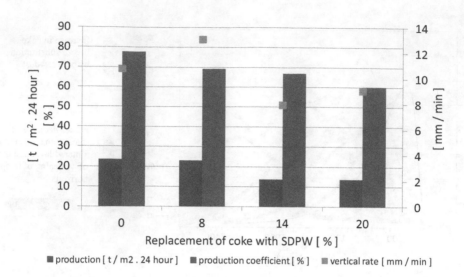

Fig. 154 Production parameters of agglomeration process for coke replacement with pine sawdust

Foreign studies determined that residues from the food industry – nutshells, husks and pits – can also be used for agglomeration process. Coke powder was substituted by walnut shells up to 50 % of the substitution. This biomass shows much more positive results in the preparation of agglocharge than sawdust. When nutshells are used, the sintering results can be evaluated positively in the whole range of substitution ratios. The walnut shells consist of several homogeneous portions formed by parenchymal cells. Even after finer grinding and enlargement of the reaction surface, their surface characteristics remain more or less unchanged, which was also confirmed in the granulometric adjustment of these nutshells.

Temperature profiles in measured locations of the sintered layer and the influence of coke powder substitution by nutshells on the distribution of the temperature field are shown in *Fig. 155* and *Fig. 156*. In contrast to sintering with sawdust, walnut shells show a positive effect on the heat transfer rate, which was confirmed by the temperature conditions in all cases.

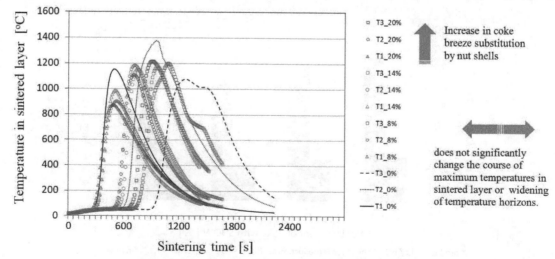

Fig. 155 Temperature curves in measured horizons of coke layer with 0–20 % of nutshells

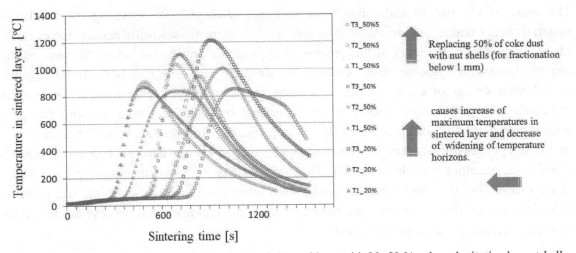

Fig. 156 Temperature curve in measured horizons of sintered layer with 20–50 % coke substitution by nutshells
(S – selected shells without dust fractions)

When analysing the production parameters of the sintering, a certain trend was observed with the replacement optimum – about 20 %, *Fig. 157*. In the optimisation step of the shell granulometry adjustment (no dust), a significant increase in the vertical sintering rate and an increase in the total specific productivity (in spite of the lowest value of the production coefficient) were observed. Also, the relatively high coke substitution by shells with separated fraction below 1 mm has brought quite positive results. Despite the general decrease in temperature peaks in the measured locations of the sintering layer in coke substitution, significant differences in strength properties of agglomerates were not observed in the case of nutshells. Even when up to 35 % of coke were replaced with shells, an increase in the strength of approx. 5 % was observed, while the abrasion deteriorated only by about 1 %.

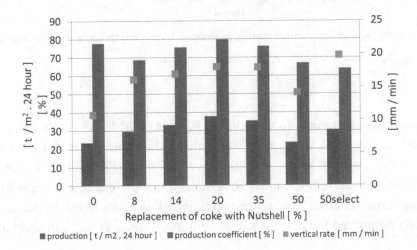

Fig. 157 Production parameters of agglomeration process for coke replacement with nutshells

The issue of the use of individual biomass types (*Tab. 37*) for sintering of ferriferous materials is not widespread in the world. Nevertheless, there are scientific papers that evaluate the technological parameters of this process. A typical example of the scientific study of the use of biomass in agglomeration is Corus, whose important objective is to investigate the partial replacement of coke powder with biomass materials in agglomeration. The papers summarise the preliminary studies that were carried out using a laboratory sintering pan, where thermal profiles of the sintered layer were evaluated. In *Fig. 158*, there is a comparison of the temperature profile in the sintered layer using coke powder and biomass materials (sunflower, hazelnut and almond) [91]. This biomass has a wider temperature profile in the sintered layer, but the maximum temperatures are lower compared to the use of coke powder. It is obvious that the maximum temperatures within the agglomeration process are lower for biomass than for the coke powder itself.

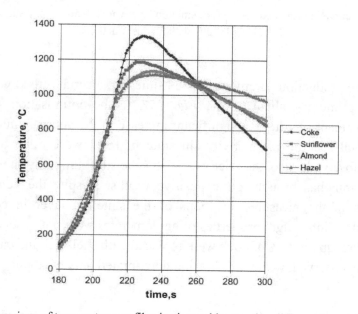

Fig. 158 Comparison of temperature profiles in sintered layer using different types of biomass [91]

Fig. 159 shows the temperatures in the sintered layer at 0–20 % coke powder substitution by sunflower seed husks [94]. It can be seen that when sunflower husks are used, the temperatures in the sintered layer are at the level of about 1200–1300 °C, i.e. there is no lowering of the maximum temperatures. The use of this type of biomass leads to a reduction in the sintering time. Due to the higher reactivity of the sunflower husks and the lower thermal capacity compared to coke, the use of sunflower husks as fuel can shorten the time of sintering.

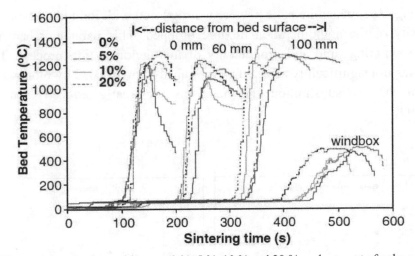

Fig. 159 Temperatures in sintered layer at 0 %, 5 %, 10 % and 20 % replacement of coke powder with sunflower husks [94]

During the pyrolytic decomposition of hydrocarbons from sunflower husks, free radicals are produced that can act as a catalyst accelerating the burning of coke. Therefore, the reduction of the sintering time when coke powder is replaced with sunflower husks can lead to an increase in the agglomeration process productivity.

Based on experiments with using 75 % of coke powder and 25 % of sunflower husks (*Fig. 160*) (grain size classes of 4–2 mm, 2–0.85 mm, 0.85–0.6 mm and 0.6–0.2 mm), the following significant conclusions can be made [91, 94]:

- temperature profiles for different grain size classes of sunflower husks vary,
- temperature profile of sunflower husks with the grain size of 0.85–0.60 mm is the most closely related to the temperature profile of the coke powder,
- for each type of biomass, it is necessary to determine the optimal grain size class for use within the agglomeration process.

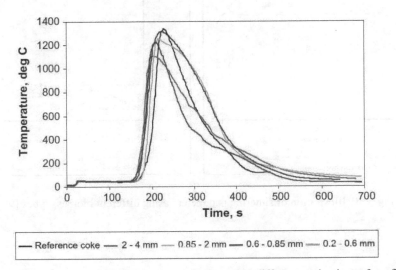

Fig. 160 Comparison of temperature profiles in sintered layer using different grain sizes of sunflower husks [91]

As shown in *Fig. 161*, when the coke powder was replaced with individual types of biomass, the bulk density of the agglomeration mixture dropped. This parameter had a significant impact on the sintering time and productivity of the agglomeration process. The yield of agglomerate was not significantly affected by the substitution of coke by biomass up to 20 %. In the case of charcoal substitution, the yield was not greatly reduced even at the 40 % replacement [99].

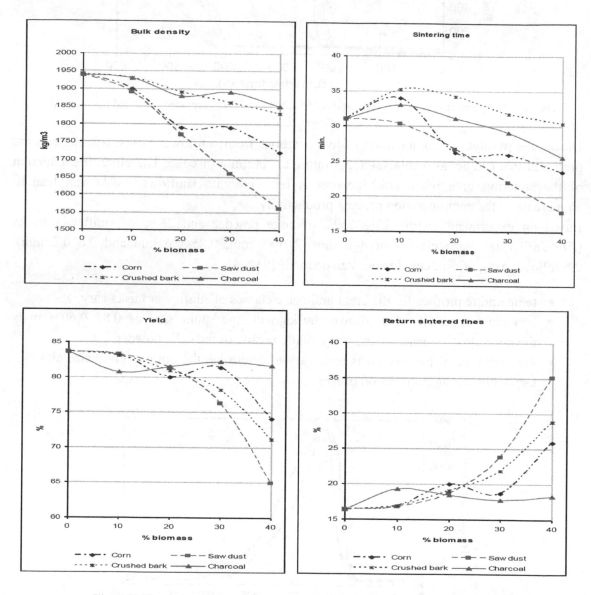

Fig. 161 Illustration of sintering parameters using different biomass types [99]

In other foreign studies, different types of biomass were tested as a substitute for coke powder in the sintering of ferriferous materials. Pursuant to the results of these studies, the following conclusions can be made (except for charcoal) [33, 39, 79, 84, 85, 113, 127–129]:

- It is possible to substitute about 10–15 % of coke powder by individual types of biomass in the agglomeration.

- Combustion of plant biomass mainly depends on the carbon structure of the cellulose, hemicellulose and lignin, which differ from the amorphous carbon in coke (coke powder).

- The maximum temperatures in the agglomeration process are lower with biomass than with actual coke powder, and the temperature rise at different levels of sintered layer occurs earlier.

- Biomass fuels can burn more quickly than coke powder due to their high porosity and large interface area, while there is a significant increase in the vertical speed of sintering. The rapid combustion of biomass reduces the maximum temperature and shortens the holding time at the high temperature, which reduces the yield and drum index.

- FeO content in the agglomerate is lower when biomass is used and significantly decreases with the more extensive biomass substitution.

- Lower temperatures in the sintered layer observed with the addition of biomass can also be attributed to the condensation of semi-volatile and volatile organic compounds. These compounds can eventually be converted into a phase similar to ash and reduce the heat transfer in the direction of burning.

The newest results of studies of coke powder substitution with biomass for the production of Fe agglomerate are summarized below, in *Tab. 42* [26, 112, 130-137].

Tab. 42 Biomass applicability in various fields - the newest findings [130]

Objective	Findings	Year, author
To study the applicability of biomass in iron and steel industry based on future challenges and opportunities.	Review is done to upgrade and analyse the opportunities and barriers of biomass adaptation in industry. Different ways of biomass utilization is discussed. Utilization of bio-char with fixed carbon > 90% and size 1–5 mm is found suitable that can replace up to 60% of coke in sintering operations.	2016 Ramchandra [131]
Review - conversion of micro-algal biomass in biofuels through thermochemical conversion.	Micro-algal biomass are characterized because it's a valuable solution for green-house gas emissions reduction and later on thermo-chemical conversion techniques are discussed to convert micro-algal biomass intobiofuel.	2015 Chen [132]
Review - compositional and property comparison between coal and biomass and their limitations.	Advantages and disadvantages are provided for biomass, and various compositional and property based points are listed.	2015 Vassilev [133]
To study fuel combustion modelling in iron ore sintering process.	Various models are discussed like coke combustion model, fuel reactivity model, coke size distribution model and studied the coke combustion behaviour with flame front speed in sintering process.	2015 Zhao [134]
Assessment of coal and biomass based on the operational and environmental impact over iron ore sinter production.	Charcoal is used as a renewable fuel with replacing 0, 10, 25, 50, 100%) of coke with charcoal and sintering time, temperature, sinter quality and GHG emissions are studied and concluded that with 50% charcoal participation results are found promising.	2015 Abreu [135]
To measure the temperature of biomass based iron ore sintering process using thermo-vision.	Thermo-vision pictures are provided of during sintering process. Results are found that sintering temperature is found lower than that of coke. Sintering rate is also found varied than the calculated one. Thermal variations are also found between the surface and inner layers (measured by thermocouple).	2014 Legemza [112]
Flue gas recirculation in iron ore sintering operations using bio-char as fuel together.	Circulating flue gas and replacing 40% of coke with bio-char resulted in increased thermal utilizing efficiency and increased consistency between heat front and flame front.	2014 Min [136]
To study the applications of biomass as a source of energy in iron ore agglomeration process.	Selected biomass are analysed chemically and microscopic studies are also performed. Structure and chemical composition of biomass and bio-ash are also discussed.	2014 Fröhlichová [26]
To substitute coke breeze by charcoal in iron ore sintering process.	Results are found negative on even 100% replacement of coke breeze with charcoal and also increasing the fuel input from 3.62% to 4.17%.	2013 Lu [137]

12.3.5. Impact of biomass on the quality of Fe agglomerate

Experimental sintering using individual biomass types was performed to verify the impact of their addition on quality indicators of agglomerate production as well. The results of laboratory experiments have brought valuable insights into the impact of the addition of specific biomass types to the properties of the produced agglomerates. The chemical composition of agglomerates produced using different biomass types is shown in *Fig. 162*. The content of Fe_{TOT} in these agglomerates does not change considerably compared to the standard (reference) agglomerate AR and is within the interval of about 51–53 %. Fe_2O_3 content in the agglomerates produced using biomass increases. The highest levels of Fe_2O_3 are found in agglomerates that were produced at 14 % coke powder replacement. The FeO content slightly increased for agglomerates where charcoal replaced the coke powder. Yet it declined sharply for agglomerates made with sawdust replacing the coke powder. These findings are also in accordance with the thermodynamic study where the sintering using charcoal produces higher amounts of CO $_{(g)}$ and H_2 $_{(g)}$ in the gas phase, resulting in a higher FeO content in the resulting agglomerate. During sintering with sawdust, a higher amount of H_2O $_{(g)}$ is present in the gas phase. Due to the lower heating value of sawdust, the temperature does not reach levels of 1200–1300 °C (as with the use of coke powder), which is also reflected in the higher Fe_2O_3 content in the resulting agglomerate.

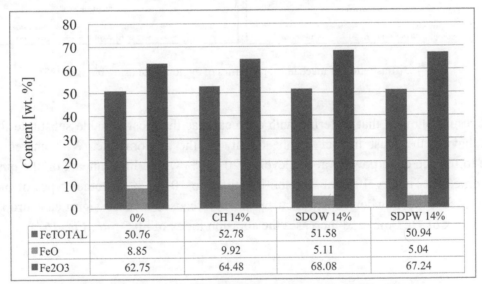

	0%	CH 14%	SDOW 14%	SDPW 14%
■ FeTOTAL	50.76	52.78	51.58	50.94
■ FeO	8.85	9.92	5.11	5.04
■ Fe2O3	62.75	64.48	68.08	67.24

Fig. 162 Chemical composition of agglomerates produced using biomass
0 % – no biomass, CH 14 % – 14 % charcoal,
SDOW 14 % – 14 % oak sawdust, SDPW 14 % – 14 % pine sawdust

The quality of the produced agglomerate was influenced by various factors, including the type and amount of used biomass, *Fig. 163*. The substitution of coke by charcoal has the least impact on the mechanical properties of the agglomerate, while the influence of sawdust was the most pronounced [99]. The strength index of +6.3 mm did not decrease up to 20 % of coke powder substitution by any biomass used. In the case of charcoal, the strength index of +6.3 mm did not change significantly even at 40 % of coke powder replacement. The most significant drop in the strength index of +6.3 mm was recorded above 20 % substitution of coke powder by sawdust. The abrasion index of 0.5 mm did not increase up to 20 % of coke powder substitution by any biomass used. However, it increased significantly above this level for all biomass types except for charcoal. The most significant increase in the abrasion index of −0.5 mm was recorded above the 20 % substitution of coke powder by sawdust.

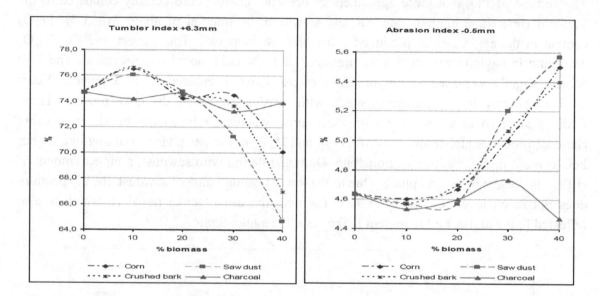

Fig. 163 Illustration of mechanical properties of agglomerate using biomass [99]

The test results showed that sintering with charcoal was the most suitable substitute for coke powder, having the least impact on the quality of the agglomerate. With charcoal, it is possible to replace coke powder at the level of 20–50 % with a little impact on the quality of the agglomerate. The level of substitution is limited to 10–20 % for other types of biomass. The increased coke powder substitution by sawdust causes a more significant drop in the strength index of +6.3 mm and increases the abrasion index of −0.5 mm, *Fig. 164*.

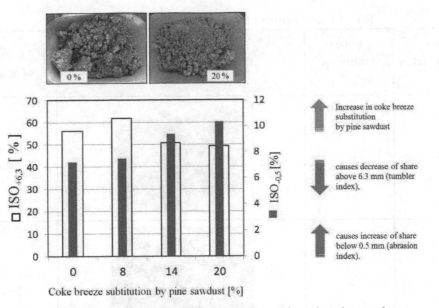

Fig. 164 Illustration of agglomerate strength properties using pine sawdust

In the case of the granulometric composition optimisation for some biomass types – e.g. nutshell, it is possible to achieve the positive effects with the replacement of coke powder, *Fig. 165*.

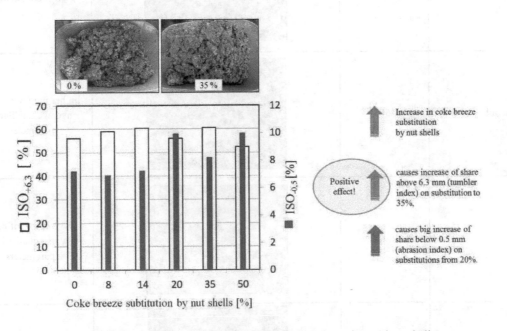

Fig. 165 Illustration of agglomerate strength properties using walnut shells

Tab. 43 gives characteristics of certain agglomerates produced with biomass substitution of coke powder. It can be seen that using charcoal and nutshells, standard quality parameters are achieved even with the higher coke powder substitution (20–44 %).

Tab. 43 Characteristics of agglomerates produced using biomass

Type of fuel	Coke substitution [%]	Photograph of agglomerate	Characteristics of agglomerate
Coke powder	0		standard qualitative parameters
Charcoal	20		standard qualitative parameters
Charcoal	44		higher volume of melt, excellent strength
Charcoal	50		higher inhomogeneity, average qualitative parameters
Charcoal	86		low volume of melt, worse qualitative parameters
Oak sawdust	14		standard qualitative parameters
Continued on next page			

Oak sawdust	44		low volume of melt, unacceptable qualitative parameters
Pine sawdust	8		standard qualitative parameters
Pine sawdust	20		low volume of melt, unacceptable qualitative parameters
Nutshells	8		standard qualitative parameters
Nutshells	20		standard qualitative parameters
Nutshells	50		standard qualitative parameters

Along with the chemical, physical and mechanical properties evaluation, the agglomerate is also evaluated in terms of its internal structure, the bond between individual components, and the arrangement of individual elements. The structure of standard agglomerates is very complex due to the polycomponent charge and many technological factors affecting the process of sintering. Agglomerates have the potential for significant inhomogeneity in their phase composition. In general, the agglomerate consists of four major phases: iron oxides (40–70 wt%), ferrites (20–50 wt%, most of which are complex compounds of calcium silicates and SFCA aluminium), glass (< 10 wt%) and calcium silicates (< 10 wt%). In the previous text of this monograph, it has already been stated that the activity of iron ions (Fe^{2+} and Fe^{3+}), calcium (Ca^{2+}), silicon (SiO_4^-) and oxygen (O^{2-}) has the crucial influence on the structure of the agglomerate. This activity will be significantly affected not only by the amount of fuel but also the type of fuel. Fuel in the form of biomass contains considerably more volatile substances, and biomass ash differs from the ash of the coke powder. This may be reflected in the agglomerate structure by increased porosity and cracks.

Porosity is evaluated within an internal analysis of agglomerates, for example, microscopically in an optical microscope. Pore size and percentage of porosity are evaluated using image analysis programs. The principle of porosity measurement is based on measuring binary 2D objects. Within the analysis of agglomerates produced by replacing the coke powder with different types of biomass, it was found that the standard agglomerate had the lowest porosity. The porosity of this agglomerate was 14 %, and the pore diameter was 127 μm, *Fig. 166*.

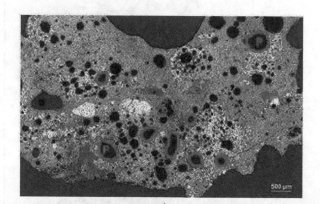

a)

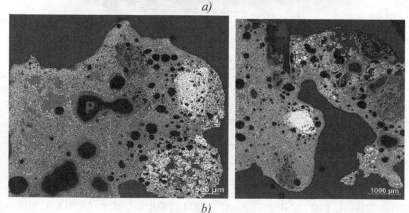

b)

Continued on next page

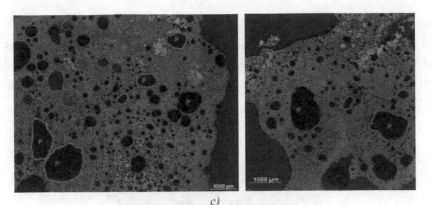

c)

Fig. 166 Pore pattern in agglomerates
a) standard agglomerate,
b) agglomerates with substitution of coke by charcoal,
c) agglomerates with substitution of coke by sawdust

In agglomerates with the substitution of coke powder by charcoal, the porosity in the range from 20 to 25 % was determined, and the pore diameter was 117–179 µm. The increased pore size for agglomerates with the substitution of coke powder by charcoal is likely to be related to the transformation of hematite into magnetite within the ongoing oxidation and dissociation processes. In agglomerates produced with the 40 % coke substitution by charcoal, the coalescence of large pores was observed, *Fig. 166b*.

The increased porosity in agglomerates with the substitution of coke powder by charcoal may be influenced by the porosity of charcoal itself. Compared to coke, charcoal has a higher porosity, and its pores are of a smaller size, *Fig. 167a*. There are mainly the larger pores in coke that are created in the process of coal carbonisation during the production of coke, *Fig. 167b*.

The maximum values within the analysed pore diameter intervals were found in agglomerates with the substitution of coke powder by sawdust. In this case, the pore diameter was 83–221 µm at the determined porosity of 17–22 %, *Fig. 166c*.

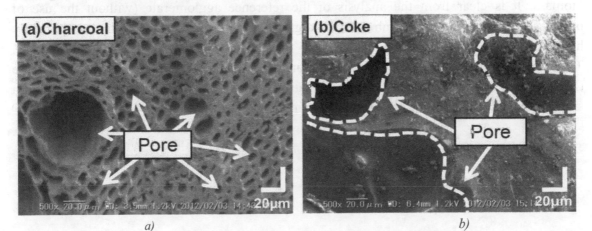

a)   b)

Fig. 167 Pore pattern in charcoal (a) and coke (b)

Fig. 168 shows the structure of the agglomerate was produced with a certain amount of biomass (20 % coke powder substitution by sunflower husks). Sunflower husks are characterised by the high content of alkaline compounds found in their ash, *Tab. 35*. The produced agglomerate was characterised by the presence of sodium chloride (NaCl) and potassium chloride (KCl) near the pores.

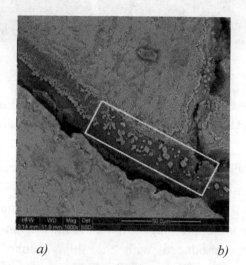

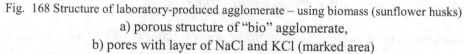

a)   b)

Fig. 168 Structure of laboratory-produced agglomerate – using biomass (sunflower husks)
a) porous structure of "bio" agglomerate,
b) pores with layer of NaCl and KCl (marked area)

With regard to the evaluation and study of the qualitative composition of agglomerates produced with biomass, it is important to know their structure also in places where fuel burns up. By analysing these parts of the agglomerate, the pores (as a result of fuel burning) can be found and the composition of grains containing unburned fuel can be determined [114]. The structure of the analysed parts changes in the case of coke powder replacement with biomass. It is clear from the analysis of the reference agglomerate (without the use of biomass) that the residues of unburned coke grains were identified mainly by exceptionally high intensities of aluminium in the EDS spectrum typical for this fuel, *Fig. 169*. The immediate surroundings consisted primarily of magnetite Fe_3O_4, calcium magnetite $(Fe,Ca)Fe_2O_4$, and two to three-phase matrix composed of calcium and aluminium silicoferrites SFCA, larnite Ca_2SiO_4 together with magnetite.

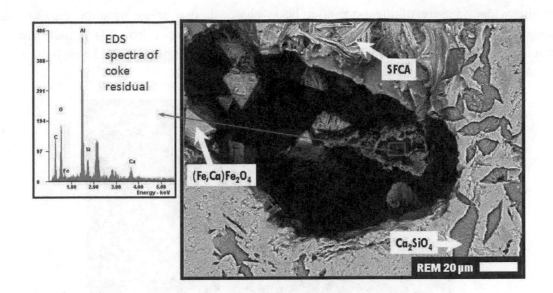

Fig. 169 Pore after combustion of coke grain [114]

The EDS spectrum shows a strong intensity of calcium for charcoal residues, which is prevalent in ash in the form of CaO, **Fig. 170**. The results of the analysis of various types of biomass ash show that charcoal ash has a strong basic character, and the CaO content is about 30–40 %, **Tab. 39**. Burnt grains of charcoal leave a dense network of irregular crystalline lime in the matrix.

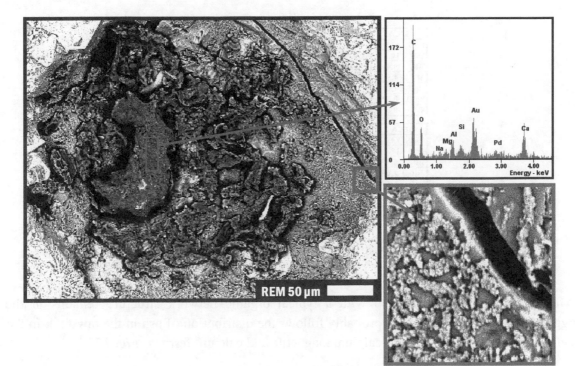

Fig. 170 Pore after incomplete combustion of charcoal [114]

In the wider vicinity of the pore after burnt charcoal, two areas with the regular phase composition formed after the solidification of iron-calcium or iron oxide melt are well discernible, *Fig. 171*. The boundary of both zones meets in the pore, around which there is no border indicating the impact of fuel on the phase composition.

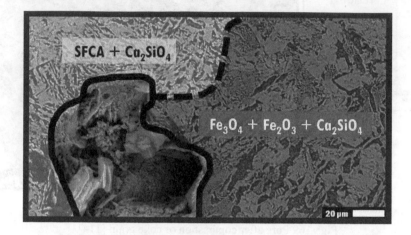

Fig. 171 Vicinity of pore after combustion of charcoal [114]

A part of ash after the combustion of fuel is assimilated by the surrounding melt and contributes to the formation of mineral phases. Silicoferrites of calcium and aluminium, which were close to the burning charcoal grains, contain the oxides of sodium and potassium in addition to the basic components, *Fig. 172*. In particular, K_2O is rarer in other phases.

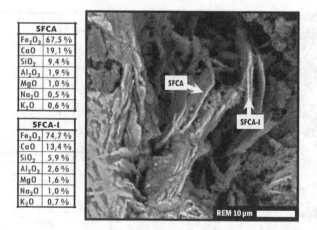

Fig. 172 Distribution of ash components in minerals [114]

After burning of nutshells, ash with a typical organic network structure remains in the agglomerate. This arrangement probably follows the distribution of ash in the raw fuel. In the vicinity, there is mainly larnite, calcium magnetite and calcium ferrites, *Fig. 173*.

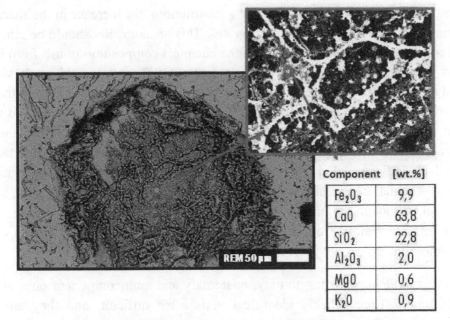

Component	[wt.%]
Fe_2O_3	9,9
CaO	63,8
SiO_2	22,8
Al_2O_3	2,0
MgO	0,6
K_2O	0,9

Fig. 173 Ash after combustion of nutshell [114]

The phase composition of selected agglomerates based on X-ray diffraction analysis is qualitatively comparable, while the differences are observable in the quantity of individual phases. In **Tab. 44**, phases are arranged in groups by their type [114].

Tab. 44 Phase composition comparison of bio agglomerates [114]

Type	Phase	AR *		CH 44 %		CH 86 %		NS 35 %		NS 50 %	
		[%]	Σ	[%]	Σ	[%]	Σ	[%]	Σ	[%]	Σ
iron oxides	Fe_2O_3	36.34		20.77		19.21		19.21		19.89	
	Fe_3O_4	23.81	**60.51**	32.43	**53.69**	34.29	**50.66**	34.29	**53.86**	19.29	**39.57**
	$Fe_{1-y}O$	0.36		0.49		0.36		0.36		0.39	
silicates	Ca_2SiO_4	6.60		7.22		10.72		10.72		10.19	
	$CaFeSi_2O_6$	6.49	**14.16**	8.34	**16.96**	6.31	**14.06**	6.31	**19.52**	7.40	**19.67**
	$CaSiO_3$	1.07		1.40		2.49		2.49		2.08	
calcium ferrites	$Ca_2Fe_2O_5$	3.07		5.32		6.91		6.91		6.56	
	$Ca_4Fe_9O_{17}$	2.35		3.91		3.81		3.81		4.72	
	$CaFe_5O_7$	1.70	**13.25**	3.21	**21.83**	6.11	**24.16**	6.11	**22.35**	7.53	**36.15**
	$Ca_2Fe_{22}O_{33}$	5.12		8.79		5.00		5.00		16.34	
	$Ca_2(Al.Fe)_2O_5$	1.01		0.60		0.52		0.52		1.00	
non-assimilated non-ore phases	CaO	-		-		1.85		1.85		2.45	
	$CaMg(CO_3)_2$	0.97	**11.23**	0.76	**6.54**	1.15	**8.57**	1.15	**4.29**	0.06	**4.60**
	SiO_2	10.26		5.78		1.29		1.29		2.09	

* AR reference agglomerate without biomass

CH 44 (86) % – agglomerate with 44 (86) % of coke powder substituted by charcoal

NS 35 (50) % – agglomerate with 35 (50) % of coke powder substituted by nutshells

Compared to standard agglomerate AR (0 % substitution), the increase in the share of calcium ferrites can be noticed in selected agglomerates. This phenomenon should be attributed to the changed conditions of sintering and not to the chemical composition of ash from burned fuels that differ considerably. An increase in the content of $Ca_2Fe_2O_5A$ is particularly visible, which is formed in locations of basic additive assimilation, where there is the largest amount of CaO donor. Low values are seen only in samples of reference agglomerate (AR) and agglomerate produced with the 86 % substitution of coke by charcoal (CH 86 %), where there is also a high proportion of the residual phase from the agglomeration ore – quartz. It can be considered an indicator of the assimilation rate of the agglomeration ore, while its other component – hematite – cannot be distinguished with this method as a primary, secondary or tertiary form. Although the proportions of additional residual phase – dolomite – seem to be even, the data are distorted by the diffraction maxima of other phases, which overlap with those of dolomite.

Besides binary calcium ferrites, ternary, quaternary and multicomponent ones are present in the agglomerate. Their reliable identification is often difficult, and they are most often referred to as silicoferrites of calcium and aluminium (SFCA). As these complex compounds are not found in the diffraction database, the proportion of the calcium ferrites was determined by the addition of all analysed calcium ferrites.

The matrix of all analysed agglomerates is composed of iron oxides. During sintering, they are partially reduced, oxidised, and to a large extent also involved in the formation of other compounds, in particular, calcium ferrites, since agglomerates in question are produced with higher basicity – approximately 1.4–1.6. Extreme cases are again represented by samples AR and CH 86 %, where the proportion of hematite is significantly higher. The reason is the presence of primary hematite in unreacted pieces of agglomeration ore, as evidenced by the high amounts of quartz. The sample NS 50 % (50 % substitution of coke by nutshells) had the lowest iron oxide content. In this case, despite the high rate of assimilation, the share of magnetite did not reach 30 %. This deficit has been caused by the stabilisation of $FeO.Fe_2O_3$ in the structure of ternary calcium ferrites $Ca_2Fe_{22}O_{33}$.

The microstructure of the standard agglomerate mainly consists of primary magnetite and hematite, *Fig. 174*. Magnetite is formed by partially molten grains with a fine crystalline structure. Geometric morphology with tetragonal and triangular particles prevails. Hematite grains are dimensionally larger than grains of magnetite and have a plate-like rhombohedral form. Hematite can be present in the agglomerates in three basic morphological forms. The primary hematite is the unreacted portion of hematite ore. The secondary hematite has a characteristic polycrystalline structure with internal pores, and the tertiary hematite was formed by reoxidation of magnetite. Forms of silicoferrites of calcium and aluminium – SFCA and calcium ferrites – are also visible in the microstructure to a small extent. The main component of SFCA is a solid solution of $CaO.2Fe_2O_3$ with SiO_2 and Al_2O_3 compounds. Their occurrence has an increasing tendency with the rising basicity of agglomerate. The higher content of calcium ferrites and their effective allocation in the microstructure improves the strength of the agglomerate. Calcium ferrites have a prismatic, rectangular or plate-like shape.

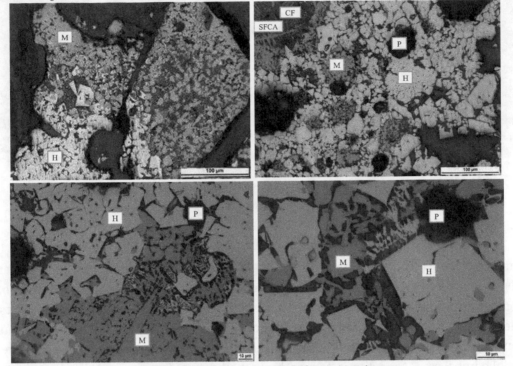

Fig. 174 Microstructure of standard (reference) agglomerate

M – magnetite, H – hematite, SFCA – silicoferrites of calcium and aluminium, CF – calcium ferrites, P – pore

The microstructure of the agglomerate with the 14 % substitution of coke powder by charcoal is mainly constituted by primary and secondary magnetite and hematite, *Fig. 175*. Magnetite is formed by partially melted grains having the fine crystalline structure. In this case, geometric morphology with quadratic and triangular particles is prevalent as well. The primary hematite is composed of unreacted parts of hematite ore, and the secondary hematite has a structure with internal pores. The hematite grains are dimensionally larger than grains of magnetite. The microstructure is visible and forms calcium silicates, which cause a reduction in the strength of the produced agglomerates.

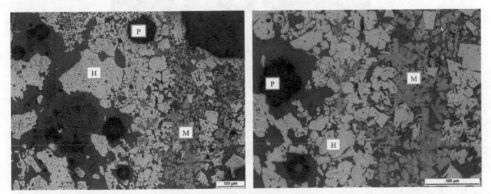

Continued on next page

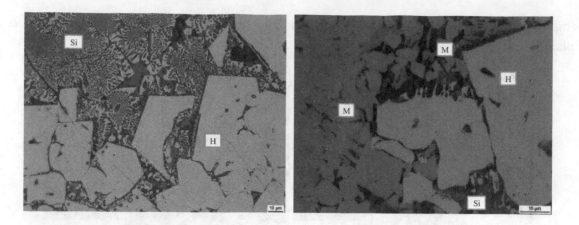

Fig. 175 Microstructure of agglomerate with 14 % substitution of coke powder by charcoal
M – magnetite, H – hematite, Si – silicates, P – pores

The microstructure of agglomerate with the 44 % substitution of coke powder by charcoal consists mainly of primary and secondary magnetite, *Fig. 176*. In the microstructure of the agglomerates produced with charcoal, the tertiary hematite is also present near the edges and in the vicinity of pores, which probably locally oxidised during agglomerate cooling. Hematite grains are dimensionally larger than grains of magnetite and have a plate-like rhombohedral form. Magnetite is formed by partially melted grains with the fine crystalline structure. Geometric morphology with tetragonal and triangular particles prevails. Forms of calcium silicates are also visible in the microstructure, and silicoferrites of calcium and aluminium – SFCA – are significant as well. Calcium ferrites are also visible in the microstructure of agglomerate with the 44 % substitution of coke powder by charcoal to a small extent. The calcium ferrite phase in agglomerate with charcoal reacted with magnetite grains and partially dissolved them. The evidence is fine dendrites based on secondary magnetite crystallised from the calcium ferritic melt.

Continued on next page

Fig. 176 Microstructure of agglomerate with 44 % substitution of coke powder by charcoal
M – magnetite, H – hematite, SFCA – silicoferrites of calcium and aluminium, CF – calcium ferrite,
P – pores

The microstructure of agglomerate with the 8 % (or 14 %) substitution of coke powder by oak sawdust is mainly constituted by hematite and magnetite, *Fig. 177, 178*. In the agglomerate microstructure, an area of an unsintered surface is visible. Shiny white areas in the microstructure of the unsintered mixture are lime. The microstructure of agglomerate with the 20 % substitution of coke powder by oak sawdust is mainly formed by primary and secondary magnetite, *Fig. 179*. In the microstructure of the agglomerate, there is also a visible area of the unsintered surface with shiny white forms, which have been identified as lime. Forms of calcium silicates are also visible in the microstructure, and silicoferrites of calcium and aluminium – SFCA – are significant as well.

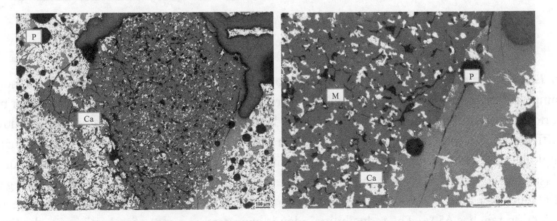

Fig. 177 Microstructure of agglomerate with 8 % substitution of coke powder by oak sawdust
M – magnetite, Ca – lime, P – pores

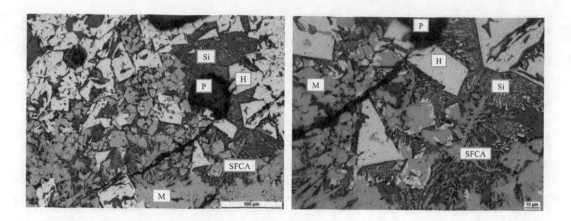

Fig. 178 Microstructure of agglomerate with 14 % substitution of coke powder by oak sawdust
M – magnetite, H – hematite, SFCA – silicoferrites of calcium and aluminium, Si – silicates, P – pores

Fig. 179 Microstructure of agglomerate with 20 % substitution of coke powder by oak sawdust
M – magnetite, CF – calcium ferrite, Ca – lime, Si – silicates, P – pores

In *Tab. 45*, the reducibility of agglomerates produced using biomass is given. For comparison, samples of agglomerates were used, which represented higher substitutions of coke powder by various types of biomass. High-temperature reducibility test evaluates to what extent and how difficult it can be to remove oxygen from the ferriferous materials (ores, pellets and agglomerates) when they are reduced under similar conditions as in the reduction zone of a blast furnace [123]. The test sample is reduced isothermally at 950 °C using a reducing gas (CO and N_2), and in certain time intervals, its weight is determined until the degree of reduction reaches 65 %. The reducibility index is expressed as a rate of reduction, i.e. the time needed to achieve the 60 % degree of reduction. The rate of reduction (dR/dt ratio) is the higher, the less time is required to achieve the 60 % degree of reduction. As it follows from *Tab. 45*, the agglomerate with the 86 % substitution of coke powder by charcoal reached the highest value of reducibility (the highest dR/dt ratio). The agglomerate with the 20 % substitution of coke powder by nutshells has the lowest value of reducibility. Agglomerates with coke powder substituted by charcoal and sawdust have a similar reducibility to the reference agglomerate without any fuel substitution.

Tab. 45 Reducibility of agglomerates produced using biomass

Reducibility	Agglomerate	Fe TOT [%]	FeO [%]	dR/dt [%/min]	Photograph
	AR 0 %	50.94	8.35	**1.15**	
	CH 86 %	51.05	5.32	**1.19**	
	SDOW 20 %	49.53	6.33	**1.19**	
	SDPW 20 %	50.25	4.10	**1.15**	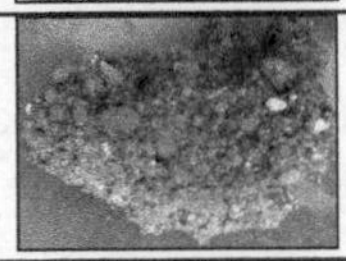
	NS 20 %	49.74	6.76	**0.94**	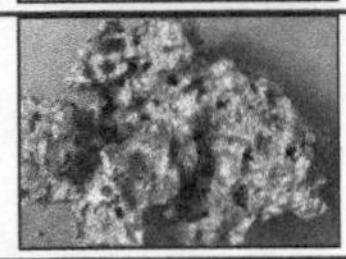

Legend:

AR 0 % – no biomass
CH 86 % – 86 % charcoal
SDOW 20 % – 20 % oak sawdust
SDPW 20 % – 20 % pine sawdust
NS 20 % – 20 % nutshells

12.4. Ecological aspects of the use of biomass in the sintering process

The current global strategies include investing in new technologies but also the development of new products to reduce their environmental impact. In this consideration, the use of economic fuels from local production and industrial processing is a good approach to the solution of environmental issues. The most important emissions into the atmosphere in the context of the use of biomass in the agglomeration process include particulate matter, sulfur oxides, nitrogen oxides, carbon monoxide and carbon dioxide.

The most important representatives of particulates include dust emitted during combustion of different types of biomass. It arises in most cases from mineral fractions of carbonaceous fuels. A small proportion of dust may also contain very small particles formed by the condensation of compounds, which gradually evaporate from the suction system of the flue gas during the sintering process. Carbon monoxide appears in the agglomeration process always as the product of incomplete combustion of both fossil fuels (e.g. coke powder) and biomass. Carbon dioxide is the main reaction product of the combustion of all carbon fuels, and its emissions are directly proportional to the carbon content of these fuels. It is released into the atmosphere not only in the reaction of carbon and oxygen during the combustion of different types of biomass, but also in the combustion of carbon monoxide or organic compounds, e.g. methane to form CO_2 and H_2O.

From the point of view of the emission profile (*Fig. 180*), a difference in the amount and course of gaseous component formation was observed during sintering with coke powder substituted by charcoal. Marked peaks of CO and CO_2 at the 50 % substitution of coke can be noticed in the measured data of emission monitoring during sintering, which attain the maximum in the halftime of sintering. This change in the pattern appears to be related to higher amounts of the volatile combustible and a condensed phase. Average values of CO, CO_2, however, suggest a positive effect of coke replacement with charcoal at higher substitutions (50–86 %). For these substitutions, average values of CO and CO_2 are by about 30–40 % lower than in sintering without the replacement of coke, *Fig. 181*. The overall effect of the charcoal addition on technological and ecological parameters of sintering is mainly positive.

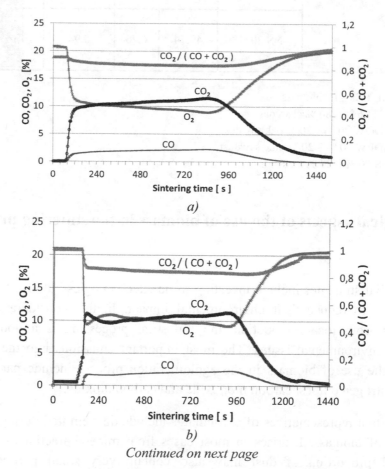

a)

b)

Continued on next page

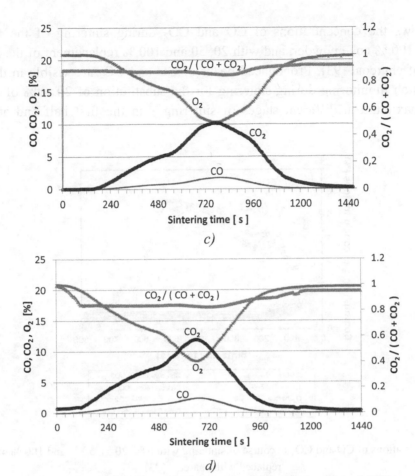

c)

d)

Fig. 180 Emission profile with % of coke substitution by charcoal (CH)
a) 0 % CH, b) 20 % CH, c) 50 % CH, d) 86 % CH

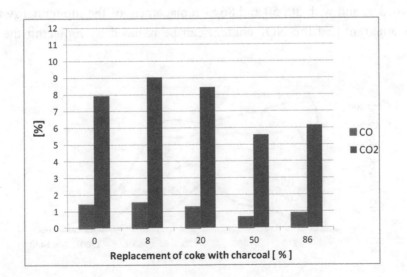

Fig. 181 Impact of coke substitution by charcoal on mean concentrations of CO and CO_2 in flue gas

Fig. 182 shows the concentrations of CO and CO_2 during sintering of the fuel mixture consisting of 100 % coke powder, and with 20, 50 and 100 % replacement of the input energy in the form of charcoal [93]. Pronounced peaks of CO and CO_2 are visible in the measured data of emission monitoring during sintering for the substitution of 20 wt% of coke, which reach their maximum at different stages of sintering – in the first half and at the end of sintering.

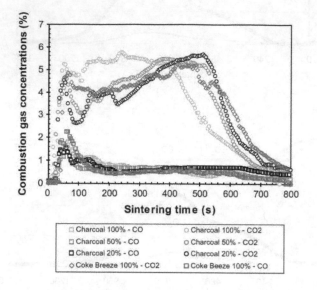

Fig. 182 Concentrations of CO and CO_2 in course of sintering with 0 %, 20 %, 50 % and 100 % of coke powder replaced with charcoal [93]

Fig. 183 shows the NO_X concentration during sintering of the fuel mixture consisting of 100 % coke powder, and with 20, 50 and 86 % replacement of the input energy in the form of charcoal. It is apparent that the NO_X content can be reduced by replacing the coke powder with charcoal.

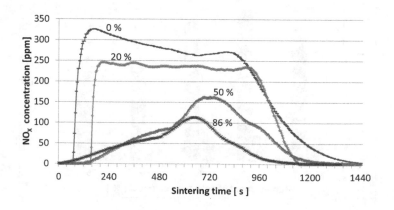

Fig. 183 NO_X concentration during sintering with 0 %, 20 %, 50 % and 86 % substitution of coke powder by charcoal

A similar dependence can be observed even when coke powder is replaced with charcoal (20–100 %) in relation to the content of NO and SO₂, *Fig. 184*. Compared to coke powder, concentrations of N and S are lower for almost any kind of biomass – including charcoal [79].

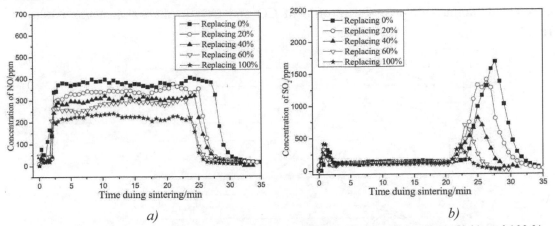

a)   b)

Fig. 184 Concentrations of NO (a) and SO₂ (b) during sintering with 0 %, 20 %, 40 %, 60 %, and 100 % substitution of coke powder by charcoal [79]

In the case of coke powder substitution by sawdust, higher concentrations of CO_2 and CO in the exhaust gas were determined at higher substitutions (20–44 %) than in the case of sintering with coke powder. The CO content was at the level of 1–3 %. The concentrations of CO and CO_2 reached the maximum values at the maximum substitution of coke powder by sawdust, *Fig. 185*.

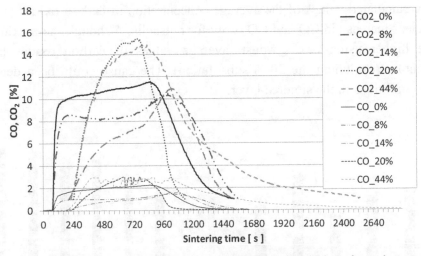

Fig. 185 Emission profile with 0–44 % substitution of coke by oak sawdust

The emission profile of sintering with the addition of selected biomass varied depending on the amount of coke replacement, as well as the type of biomass. When sintering with oak sawdust, there is a visible change in the profile curves of CO and CO_2 in terms of creating peaks in halftime of sintering, and higher concentrations of the measured peaks of those gaseous components. This change in the pattern appears to be related to higher amounts of the volatile combustible and condensed phase. In the case of oak sawdust, the ratio of $CO_2/(CO + CO_2)$ increased from 0.85 to 0.9 due to the increase of CO_2 in the flue gas. CO was at the level of 1–3 %. The concentration of CO and CO_2 reached the maximum value at the maximum substitution of the coke powder by sawdust. The effect of coke substitution by the selected biomass on emission profile of CO and CO_2 in the exhaust gas is shown in *Fig. 186*.

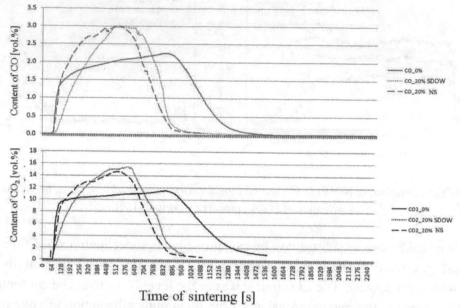

Fig. 186 Effect of selected biomass substitutes on emission profile of CO and CO_2 in flue gas
SDOW – oak sawdust, NS – nutshells

Based on the calculated values of the average concentration of carbon oxides (*Fig. 187*), we can see the clear increase in CO and CO_2 only for nutshells. Although the maximum attained values were higher using oak sawdust, average values of carbon oxides had a rather decreasing nature (excluding the 20 % substitution). The cause could be a deterioration of the sintering conditions in the sintered layer.

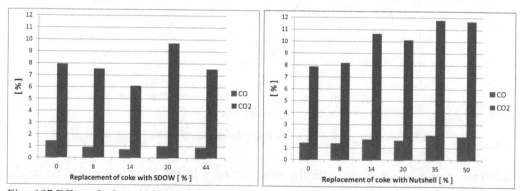

Fig. 187 Effect of selected biomass substitutes on average concentrations of CO and CO_2 in flue gas

Fig. 188 shows the SO_X concentration during sintering with the fuel mixture with 100 % coke powder and 25 % of energy input in the form of different types of biomass (sunflower, hazelnut and almond). Lower levels of SO_X emissions were observed for all sintering using various types of biomass. This finding is in accordance with the lower sulfur content in the biomass material compared to the coke powder.

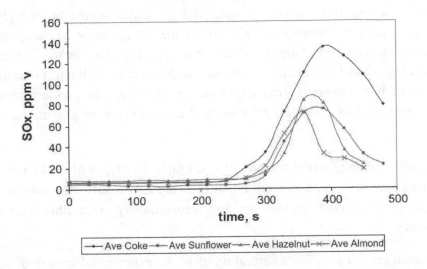

Fig. 188 SO_X concentrations during sintering using biomass

The agglomeration process is a significant source of persistent organic pollutants (POPs) such as PCDDs and PCDFs, which are considered to be dioxins. Previous studies have shown that PCDD/F emissions are associated with the burning of carbon. Production of PCDD/Fs in sintering by synthesis mechanism occurs at the temperature of 250–450 °C. According to the new hypothesis, the PCDD/Fs are formed from chemically unrelated substances such as PVC, and other chlorohydrocarbons, or from the organic aromatic compounds such as cellulose. Combustion of plant biomass depends mainly on the carbon structure of the cellulose, hemicellulose and lignin, which are different from the amorphous structure of the carbon in coke. Due to this different structure, it is necessary to evaluate the emissions of toxic organic pollutants such as PCDD/Fs and PAHs. PCDD/Fs are also formed from pentachlorophenol (PCP), which can enter biomass from plant spraying.

12.5. Economic aspects of the biomass use

Biomass was for a long time before the industrial revolution a sole or strongly dominant source of energy for the humankind. Its share of the total energy consumption began to decline sharply after the onset of the era of coal, but especially after the massive expansion of machinery burning oil and gas mainly in industry and transport. In the last decade, however, the importance of biomass energy has increased significantly. The pressure to increase the production of biomass fuel and energy thereof will continue in a long-term, mainly for three reasons:

- Rapidly declining world reserves of fossil fuels. Existence of the current civilisation depends on their consumption. Fossil fuels are historically unprecedented and most concentrated, most technically and economically accessible primary energy sources.

- Gradual reaching of finite natural limits of the exponential growth of consumption by the consumer civilisation. If the increasing consumption of energy is the engine of economic growth, maintaining the current economic model depends on the ability of humankind to replace today's dominant finite fossil fuels by others as quickly as possible. Biomass often offers a seemingly simple and quick solution.

- More and more substantial facts on the rate and status of global climate change and projections of its future economic, social and environmental consequences. The growing concentration of greenhouse gases in the atmosphere is the decisive factor that has caused this irreversible change. They are unnaturally concentrated in the atmosphere mainly due to burning of fossil fuels since the beginning of the industrial revolution. It is, therefore, necessary to replace the dominant fossil energy sources by alternative ones as soon as possible. Also, in this case, the biomass appears as the simple and fast solution.

The growing demand for biomass for energy use, in addition to the growth of primary biomass prices, has also other repercussions: the intensification of wood production or production of energy crops, and expansion of logging into the sites with previously restricted or prohibited commercial exploitation. All these trends are inconsistent with the principles of sustainable and efficient use of energy biomass as a renewable resource. The uncontrolled growth of biomass consumption and bad practices during the entire cycle of its energy use can gradually endanger and destroy the natural conditions for its continuous renewal or may threaten the quality of the environment.

Therefore, only the biomass, whose use does not compromise the long-term potential of regenerative and ecological stability of the site of its origin, should have the status of a sustainable renewable source.

The return on energy (EROEI – Energy Returned on Energy Invested) is an essential parameter to compare the profitability of different types of energy sources throughout their life cycle. It expresses the ratio between energy gains and energy expended to achieve it. The higher the value, the more favourable the particular fuel.

An important factor affecting the economic viability of biomass use is transport costs for the supply of biomass to the customer, or for further processing.

The basic criteria, on which decisions about the use of biomass for energy purposes should depend, are in particular analysis and balance of environmental and economic perspective. While the environmental criterion should be linked primarily to the evaluation of the impact of biomass energy use on the environment, the economic criterion should focus primarily on the assessment of the possibilities of economically efficient processing of biomass. The analysis of the costs and competitiveness of using various types of biomass shows that in the area of plant biomass, it appears advantageous to use waste biomass, and that specifically grown energy crops will be difficult to promote in terms of the resulting economy without subsidies for the fuel market. Energy use of biomass as a biofuel may economically mean both financial savings for the disposal of waste products and the possibility of financial savings in the form of replacing the traditional fuels (or other fuel) by biomass. This aspect, however, depends on the economic comparison of different types of fuel. The following can be included among the economic aspects of the biomass use for energy purposes:

- production of biomass as a fuel (depending on the choice of biomass type, specific location and processing technology, and sales),

- transport costs as a part of the logistics costs,

- investment costs for the adjustment of technology to allow processing biomass (economic estimate of the effect of applying the knowledge from laboratory research in practice is that it will be necessary to extend the existing technological nodes to achieve the implementation of a particular type of biomass in the standard process flow); that means mainly increasing the capacity for biomass storage, means of transportation, and pre-treatment of biomass in new processing facilities,

- costs of storage or ensuring reliable long-term supply of biomass.

13. Use of alternative fuels in the production of pig iron

Alternative fuels, which are currently used in the blast furnace process, include biomass, polymeric organic materials – plastics, regenerated waste oil, coal tar and granulated carbonaceous materials. Biomass for use in a blast furnace is prepared by the pyrolysis, heating in the inert atmosphere, and gasification. Most of the products of these processes is used only on a laboratory or pilot scale within the blast furnace technology.

Slow pyrolysis (carbonisation) takes place in the thermal conversion of solid biomass without air, and the product is charcoal. The calorific value of charcoal, which is used in the blast furnace process, is about 27–32 MJ/kg. The higher the temperature of the carbonisation, the harder and stronger the charcoal. Charcoal is the most common reducing agent in Brazil, where it is used mostly in smaller blast furnaces. For these furnaces, there are lower requirements for mechanical properties of reductants – mainly for strength. Charcoal, which is charged directly into blast furnaces, is produced by the high-temperature pyrolysis, i.e. above 800 °C. At lower pyrolysis temperatures (about 500–800 °C), charcoal is produced that can be a part of the charge for bio-coke production [115, 116]. Such a charge may also contain raw biomass, *Fig. 189*.

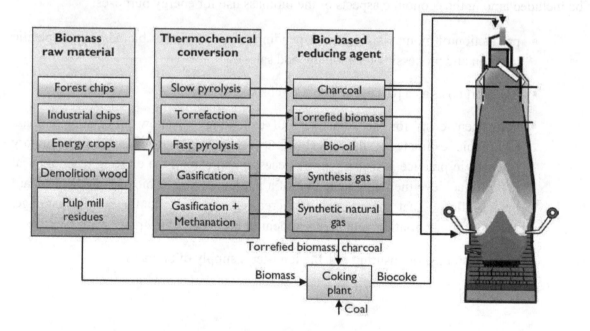

Fig. 189 Possibilities of biomass utilisation in blast furnace process [115, 116]

Heating in the inert atmosphere is another thermochemical process used for the treatment of raw biomass for the blast furnace process. At temperatures of about 200 to 350 °C, so-called refined biomass is obtained, which can be a part of the batch in bio-coke or can be injected directly through tuyeres of the BF.

So-called fast pyrolysis is one of the most advanced and promising technologies in the group of processes, which convert biomass in the form of wood and other waste materials into products of a higher energy level, such as gases, liquids and solids. The process of rapid pyrolysis is based on the rapid supply of heat into the feedstock, maintaining the required temperature in the pyrolysis reactor (about 450 to 600 °C), and short residence time of feedstock in the reaction zone (up to several seconds). Products of this process must be immediately cooled and condensed, while a large proportion of a dark brown liquid of low viscosity is formed. This liquid is called bio-oil, with the heating value of about 16 to 24 MJ/kg. By-products of the fast pyrolysis are pyrolysis coke (15 %) and pyrolysis gas (up to 40 %), which are mainly used in its own pyrolysis process for the production of heat. Bio-oil can be injected into the blast furnace as a replacement (alternative) fuel. Production of bio-oil by fast pyrolysis can be performed from any solid biofuel. It is fundamental for limiting the higher water content of the bio-oil to use pre-dried biomass with the moisture content below 10 %. Another condition is to maintain the grain size of particles below 3 mm, ensuring rapid progress of the reaction and easy separation of the components.

Another method of biomass transformation that can be used for the blast furnace process is a thermo-chemical conversion of biomass at high temperatures with limited access of air, oxygen or water vapour. By rigorous control of the temperature (800–900 °C), the content of gasification agents and the residence time of particles of biomass in the reactor (one second to tens of seconds), virtually any organic material can be turned into gas. When using atmospheric oxygen (which is the most common in the case of biomass), the resulting raw gas has a relatively low calorific value (4–8 MJ/m^3), containing about 18–20 % of CO, 18–20 % of H_2, 2–3 % CH_4, and the rest being nitrogen. Gas contains tars, phenols and particulates. The use of this gas for the blast furnace process requires refining operations. Better results are achieved with oxygen or water vapour as the gasifying agent when the gas calorific value of approximately 20–25 MJ/m^3 is obtained. Utilization of gas produced by gasification of biomass is suitable for many technological processes where such gas can replace natural gas.

In *Tab. 46*, the chemical composition and calorific value of gas that can be used in the process of pig iron production is provided. As can be seen, the calorific value of some synthesis gases made by gasification of biomass is very low and is only about 1/3 of the calorific value of natural gas [117]. Therefore, the substitution of natural gas by some alternative synthesis gases with low calorific value in the BF is complicated. The injection requires the introduction of a larger volume of synthesis gas, reducing the temperature of the BF oxidation area. Such technology would require an increase of hot air temperature and an additional injection of, e.g., coal with high calorific value.

Tab. 46 Chemical composition and calorific value of gases for production of pig iron in BF [117]

Gas	Chemical composition [%]						Calorific value
	H_2	CO	CO_2	CH_4	C_2H_{4-6}	N_2	[MJ/m^3]
Natural gas	0.8	-	0.7	88	8.5	0.9	34.5
Blast furnace gas (recycled)	15	72	3.2	-	-	10.5	10.7
Coking gas	67	6	2	20	3.5	2.5	17.5
Synthesis gas (from biomass)	40	25	21	10	2.5	1.5	14.0
Synthesis methane gas (from biomass)	26	20	35	13	3	3	12.0

Catalytic liquefaction is a process that can potentially produce better quality products with higher energy density than other thermo-chemical processes. It is a low-temperature, high-pressure thermochemical conversion process taking place at a temperature of approximately 300–350 °C and pressure of 12–20 MPa in an aqueous medium. The technology is still in experimental development stage, but due to the high quality of the final product, it may be very interesting in the future.

The biomass that can be used as an alternative source of fuel for blast furnaces may be based on waste residues from agriculture, forestry and related industries, as well as the biodegradable fraction of the industrial and municipal waste. As mentioned above, besides the industrial use of charcoal in small blast furnaces, other products of biomass processing for the blast furnace process are still only being experimentally tested and verified.

The technology for production of pig iron in a modified blast furnace – ULCOS-BF – can be classified among the most promising technologies. This process is recycling blast furnace top gas back into the blast furnace after removal of CO_2. The technology is currently being tested on a commercial blast furnace with integrated CO_2 transport and storage, *Fig. 190*. In the ULCOS-BF project, the dosage of biomass is also experimentally tested. The bulk biomass is dispensed along with ore portion of the charge from above, and fine-grain biomass is injected with coal in oxygen stream [116].

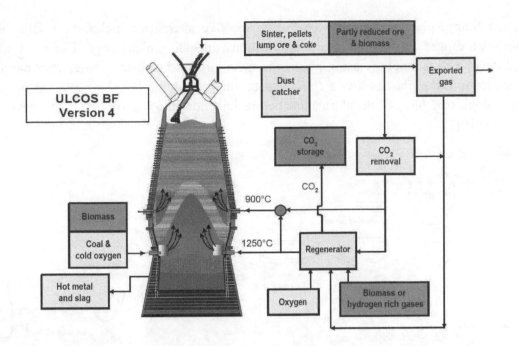

Fig. 190 Project of pig iron production in modified blast furnace – ULCOS-BF [116]

In the project of advanced DRI shaft furnace (Midrex type), mathematical models and simulations are currently being developed, in which hydrogen, produced by electrolysis of water, would be used for iron oxide reduction. This new technology has the potential to reduce CO_2 emissions by up to 80–90 % in comparison with a blast furnace. The project for Fe oxide pellets reduction with hydrogen in laboratory conditions is called Reductor. This technology has not only environmental but also technological benefits because the reduction with hydrogen is carried out at temperatures of about 800–850 °C and is kinetically faster and more efficient compared to carbon reductants (CO and C). In *Fig. 191*, microstructures of products of reduction with hydrogen at 800 °C are listed [118].

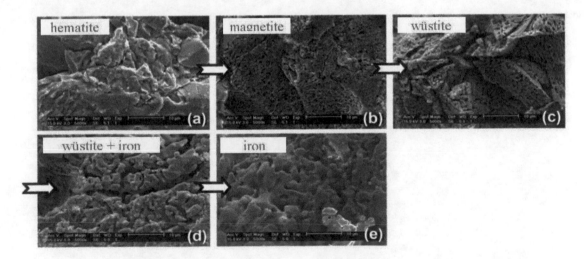

Fig. 191 Microstructure of products after reduction with hydrogen at 800 °C [118]
a) input Fe pellet b - d) reduction intermediates, e) reduced Fe

In the future, plastics may also be the prospective alternative fuel for the BF. Globally, plastics are used in many industrial applications, including metallurgy. The most advanced and developed technology using plastics in metallurgy is in Japan, where, after particle size modification, the plastics are injected into the tuyeres of the BF, *Fig. 192*. At present, the technologies of plastics dechlorination before injection into the BF are being developed in Japan [119].

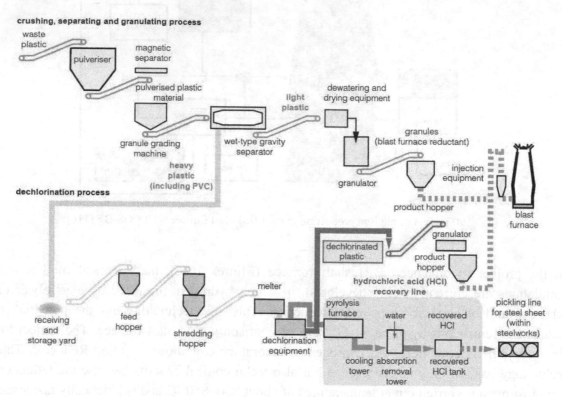

Fig. 192 Injection of modified plastics into BF tuyeres (Kobe Steel, Japan) [119]

14. Use of alternative fuels in the production of steel and ferroalloys

Alternative fuels are so far used only in a limited range within the production of steel in main steelmaking units (oxygen converter and electric arc furnace). This is because carbonaceous materials with a carbon content of more than 90 % are required in the process of steel carburising, which cannot be achieved, e.g., in the case of biomass. Carburising additives should include a minimum of volatile constituents, which again eliminates the use of biomass. For increasing the efficiency of the electric arc furnace, e.g. synthesis gas from gasification of biomass may be used in the fuel burner. Air is not suitable as a gasifying agent for this purpose, because the resulting gas has a very low calorific value (4–8 MJ/m^3). For natural gas replacement with synthesis gas, the gasification of biomass should be carried out using oxygen or water vapour, when the produced gas has the calorific value of 20 MJ/m^3. Gases from the gasification of biomass with lower calorific values can be used in auxiliary burners, which are used in systems for preheating of scrap steel for the electric arc furnace. Temperatures up to about 1000 °C are required for steel scrap heating. In that case not only auxiliary burners are used, but mainly the heat of hot gases from the production of steel in electric arc furnaces.

Today, an alternative carbonaceous fuel based on low-temperature coke is also produced, so-called fluid coke, which is also used in steel production. Fluid coke is produced from crude oil refining residues, and from carbonaceous suspensions with a high content of hydrocarbons, arising in the treatment of coal. The manufacturing process takes place at a temperature of about 565 °C in a fluidised bed reactor, where products with 92 % content of carbon, 2–4 % of volatile combustibles, 1–2 % of ash, and 2–3 % of sulfur are acquired [120]. Due to the effects of a certain proportion of the liquid phase of crude oil residues, produced fluid coke has globular shaped grains, *Fig. 193*. The globular shape of fluid coke grains is very beneficial in the reduction of FeO oxide, which is present in slag after the stage of scrap melting in the electric arc furnace. As burners of JetBox type are used for melting of steel scrap, the scrap melting stage has a strongly oxidising nature.

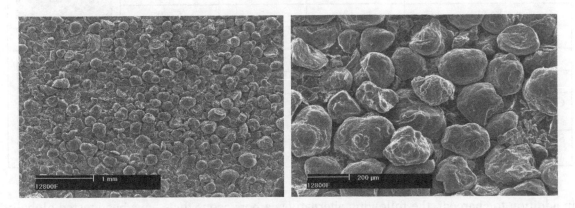

Fig. 193 Microstructure of fluid coke [120]

Since the reduction of FeO to Fe takes place at the liquid steel-molten slag interface, the elevated sulfur content in fluid coke is not detrimental to the quality of the produced steel. Within the reaction (40), the resulting carbon monoxide (CO) passes through the slag and froths it. Frothing of slag during steel production in the EAF is essential for lining protection against aggressive electric arc, which is generated on graphite electrodes.

$$\text{FeO}_{(slag)} + \text{C} = \text{Fe}_{(steel)} + \text{CO}_{(g)} \qquad \Delta H_{298} = 156.729 \text{ kJ} \qquad (40)$$

In the production of ferroalloys, various alternative reducing agents are currently being tested worldwide. For the production of siliceous ferroalloys in some countries (Brazil, India and Norway), charcoal in combination with coke is used. In Brazil, many facilities use only charcoal. Although industrial charcoal generally has lower carbon content and higher content of volatiles than metallurgical coke, it usually has lower ash content (depending on the type and location of wood). Charcoal has higher reactivity (and higher porosity) than metallurgical coke, lower thermal stability, and higher specific electrical resistance than metallurgical coke, *Tab. 47*. These properties favour charcoal over traditional reducing agents – bituminous coal coke. The disadvantage is the price of charcoal, which is about 2 to 3-times higher than pea coke in Europe, or medium coalified bituminous coal.

Tab. 47 Typical properties of charcoal and coke used in the production of ferroalloys

Properties		Industrial charcoal	High-quality charcoal	Metallurgical coke	Microstructure
Carbon	[%]	65–85	92–94	86–88	coke
Volatile matter	[%]	15–35	3.8	0.5–2	
Ash	[%]	0.4–4.0	2.0–2.5	10–12	
Analysis of ash	[%]				
SiO$_2$		5–25	23	25–55	
Fe$_2$O$_3$		1–13	5	5–40	
Al$_2$O$_3$		2–12	5	13–30	charcoal
CaO		20–40	11–30	3–6	
MgO		5–12	6	1–5	
CO$_2$ reactivity at 1060 °C [%C/s]		2.1–$2.3.10^{-2}$	2.8–$3.2.10^{-2}$	0.2–$0.5.10^{-2}$	
Specific electric resistance at 1000 °C [Ω.m]		0.014–0.023	0.015–0.030	0.003–0.008	

In addition to charcoal, the following alternative carbonaceous fuels have been tested globally in the ferroalloy plants:

- peat,

- lignite,

- wood chips,

- wood and coal briquettes,

- semi-coke,

- petroleum coke.

The share of these alternative reductants in the charge constitutes about 10–30 % replacement for conventional pea coke, and it has different impact on the technology of ferroalloy production. When the reducing agents of higher purity are used (low ash, sulfur and phosphorus), alternative carbonaceous reducing agents in most cases improve technological and economic parameters of ferroalloy production. This is due to the better physical parameters of alternative reductants – especially higher electrical resistivity and higher reactivity than that of pea coke.

15. Alternative fuels and environment

At present, it is necessary to rethink the value system on which our economic activity is built. The relationship between the humans and the nature has to assume new quality because our planet's reserves (including natural resources) are being consumed and there is an ecological imbalance. New solutions allowing to reduce the environmental burden of human activities must be found. Such solutions include all activities that will contribute to the development of technologies with the minimum power consumption or using "greener" energy. Low-energy, and at the same time, ecological technologies must be developed despite the fact energy companies will, for the sake of their profit, steadily pursue a policy of the need for considerable amounts of energy for the need of the humankind. This leads to a paradoxical situation when in developed regions (Europe, USA, Japan and Southwest Asia), there is currently an excess of energy, while in poorer regions (South America and Africa), there is an energy deficit. Yet, even in the poorer regions, there are still sufficient energy sources. However, due to the political instability and considerable poverty of a large part of the population, these sources are not used efficiently. The biggest problem is that at current rates of extraction of conventional fossil fuels, we will not have natural gas and crude oil in about 50 to 80 years and coal in about 200 to 300 years. From this perspective, it is necessary to seek and develop new alternative energy sources. In this context, renewable energy sources have the highest potential.

In earlier chapters of this publication, it has already been established that biomass in particular has the highest usable potential of all renewable energy sources for metallurgical application in the near future. There are high hopes regarding biomass (mainly woody biomass and phytomass) intended for energy and metallurgical use becoming an alternative renewable energy source and gradually replacing a part of the conventional non-renewable fossil fuels. There is a number of decisions related to biomass that is to be used either for energy purposes or is considered within the metallurgical technology, which ultimately can cause substantial economic losses. The energy market is now negatively stimulated in favour of individual investors' interests in conflict with societal needs. Often, many important facts about the acquisition and use of biomass are suppressed. For example, that the CO_2 neutrality regarding the processing of biomass will never be reached. The biomass production, its subsequent treatment (i.e. conversion into higher added value products), transport and processing are activities releasing carbon dioxide. Some operators of technologies using biomass energy point to the fact that energy recovery from biomass is not directly reducing CO_2 emissions compared to coal. Even compared to the burning of natural gas, CO_2 emissions increase by 60 % per unit of energy produced, and it deepens the climate change crisis. Moreover, in recent years, there has been a significant devastation of the environment related to the provision of biomass (primarily wood and subsequent manufacturing of charcoal) in many regions of the world (e.g. South America and Africa). In the short term, biomass can be assessed as economically favourable ecological carbonaceous fuel, but in the medium and long-term, the use of biomass may cause considerable economic, environmental and social damage.

In connection with alternative fuels, we must adopt solutions that minimise the use of raw biomass (wood or charcoal and agricultural phytomass) and maximise the use of waste biomass. The advantage of using waste biomass is that it arises as a by-product, so it is not attributed to the use of forest and agricultural land "for energy production". A sensible approach could create such conditions that raw biomass for energy and industrial (e.g. metallurgy) purposes will be grown and used only on defined and controlled areas. Presently, there are some positive examples of such solutions, where plantations of fast-growing trees and vegetation are a part of a metallurgical company (e.g. Japan, Germany and Sweden). Nevertheless, the future trend should be primarily the use of waste biomass. Through the processing of waste biomass, we can get rid of waste that would otherwise not have been used, and concurrently, contribute to the reduction of emission burden and improve the environment. Now, the reader may ask, whether we have enough waste biomass for increased replacement of traditional fossil fuels, which are currently utilised for energy and industrial use, and whether the use of such biomass will actually reduce emissions of, e.g., carbon, nitrogen and sulfur. The specific results of research task solution in the SR and worldwide show that the replacement of fossil fuel in the metallurgical sector by biomass is limited to the level of about 10–30 %. With such replacement, using certain types of biomass, emissions of oxides of carbon, nitrogen and sulfur would be reduced by 5–40 %. Is it a little or a lot? When we consider conditions of the SR, with the substitution about 10 % of coke powder by biomass in the agglomeration process, we could save about 10,000 tonnes of coke each year. Of course, this quantity of coke would have to be replaced with approximately two to three times higher amount of biomass (due to its physical properties and calorific value). In Slovakia, it is possible to gain 0.5 to 3 tonnes of waste biomass from one hectare of forest or agricultural land, which is currently not used. There are residues after harvesting of cereals, surplus hay, calamity wood, unused pieces, sawdust, etc. Obtaining, processing and effectively utilising waste biomass in the volume of tens of thousands of tonnes annually for metallurgical purposes is more a question of the future and further study. Even the authors of this publication do not consider biomass the sole and the best source of alternative carbonaceous fuel for the metallurgical sector. Nevertheless, they want to show the potential that biomass represents. The authors focused their research of biomass types (charcoal, wood sawdust and nutshells) mainly on the technological possibilities for using a particular type of biomass in the agglomeration process. Specific material and economic balance (including environmental benefits) should be implemented only after a thorough material research of various types of biomass and comprehensive technological assessment of all the important advantages and disadvantages of using a particular type of biomass. Therefore, it will be necessary to continue with the research and the search for alternative forms of carbon fuels in the conditions of the Slovak Republic. Research into alternative fuels (including biomass) takes place globally, including the sector of the metallurgy of iron, steel and ferroalloys.

In chapter 10, it was noted that a variety of solid, liquid and gaseous substances is included among the fundamental, i.e., the most strictly monitored pollutants, originating from burning or other thermal use of traditional carbon fuels. These are, e.g. particulates, carbon dioxide, carbon monoxide, nitrogen oxides, sulfur dioxide, organic gases and vapours measured as total organic carbon, polychlorinated dibenzo-p-dioxins, polychlorinated dibenzofurans, heavy metals, hydrogen fluoride, halogen compounds, unburnt hydrocarbons and non-methane volatile organic substances. Many of these problematic substances also arise during heat treatment of alternative carbonaceous fuels [121, 122]. Therefore, an issue is not only the burning of fossil fuels but also biofuels, which have different characteristics than fossil fuels, and are considered an ecological fuel. The environmental impact of emissions from the combustion of biofuels, as shown by experimental measurement and analyses, are in fact often not such as expected. Carbon dioxide and carbon monoxide are not the only products of fossil fuel combustion. They are also formed during combustion of biomass. Produced quantities of these gases vary within the metallurgical technology and thermal processing of biomass of a particular kind, but the overall effect of using biomass is mainly positive. The research shows that the use of biomass in most technological applications reduces the amount of CO_2 and CO generated.

From metallurgical plants using biomass, solid particles are carried away with the flue gas stream as particulate matter pollutants. These consist of inorganic substances (ash), organic matter (non-volatile flammable) and soot. The quantity of carried particles depends on many factors (e.g. particle size of fuel, ash content in fuel, thermodynamic and kinetic conditions of the production process, etc.). The advantage of using biomass is its significantly lower ash content. Another positive is a considerably lower proportion of heavy metals in biomass. Biomass ash generally contains a number of alkali metals (Na, Ca, Mg, K and P), which may become a component of mineral fertilisers. Fine dust particles are very dangerous to the environment and human health – especially the finest dust particles, less than 2.5 microns in size, capable of lasting up to several years in the upper atmosphere, which allows for their long-range transport. Their main source is not only the combustion of coal but also biomass (e.g. wood).

Sulfur dioxide resulting from the burning of fuels containing sulfur together with solid particles belong to the typical and most common components of the emissions polluting the environment. In view of the low sulfur content in biomass, the amount of sulfur dioxide resulting from the combustion of biofuels is significantly lower than from fossil fuels (mainly coal and coke). A similar correlation can also be used for the produced nitrogen oxides, the quantity of which is significantly lower in the case of thermal treatment of biomass.

Besides the thermal treatment of coal and coke, a potential source of polychlorinated dibenzo-p-dioxins is also equipment for heat treatment of wood or wood waste that has been modified or contains chlorinated organic compounds (glued sawdust, PVC, NH_4Cl, etc.).

Currently, there is a number of technologies being globally developed that seek to prevent or reduce the proportion of fossil carbonaceous fuels (mainly coke) in iron and steel metallurgy [124]. There are known melting-reducing and direct iron production technologies that use solely bituminous coal or natural gas for reduction. All these processes are in the minority and cannot significantly reduce carbon emissions, as carbonaceous materials are used for the reduction. Therefore, in the last 15 years, technologies have been developed that reduce the amount of CO_2 emissions by more than 50 %. For processing of secondary ferruginous materials, there are also plasma technologies being developed. The multistep reduction-melting process involves phases of pre-reduction, melting and final reduction to a liquid product with a nature of pig iron [125].

The most important of such technologies of recent years include:

1. ULCOS – BF,

2. Hisarna,

3. ULCORED,

4. ULCOWIN – ULCOLYSIS.

All these technologies are being investigated in the ULCOS project (reduction of CO_2 emissions in the iron and steel metallurgy by 50 % by 2015), which envisages the use of biomass, replacing carbon-based reductants by hydrogen from electrolysis of water, and CO_2 trapping (fixation), *Fig. 194*.

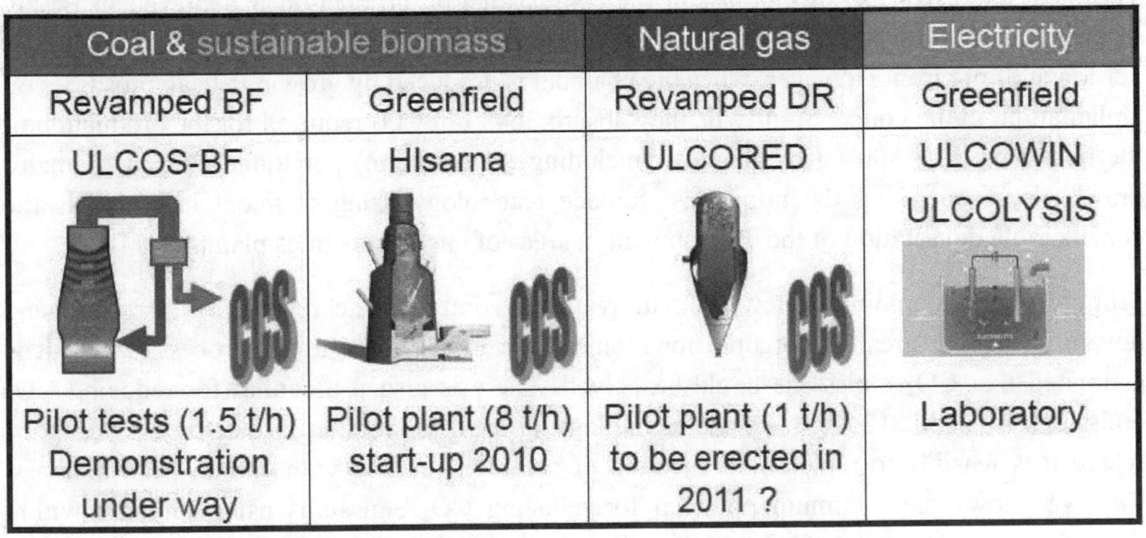

Coal & sustainable biomass		Natural gas	Electricity
Revamped BF	Greenfield	Revamped DR	Greenfield
ULCOS-BF	HIsarna	ULCORED	ULCOWIN ULCOLYSIS
Pilot tests (1.5 t/h) Demonstration under way	Pilot plant (8 t/h) start-up 2010	Pilot plant (1 t/h) to be erected in 2011 ?	Laboratory

Fig. 194 Projects to reduce CO_2 emissions within ULCOS

The ULCOS-BF process has already been mentioned above in chapter 13, "Use of alternative fuels in the production of pig iron". The technology was tested in the years 2010–2012 in an experimental blast furnace LKAB in Lulea, Sweden. Due to the positive results of experiments, the technology is now being tested in a commercial blast furnace with integrated CO_2 transport and storage. The technology is ready for use on an industrial scale.

HIsarna combines a pre-reduction reactor and melt-reduction furnace. The estimated CO_2 reduction compared to the conventional blast furnace is by 20 %. If underground CO_2 storage (CCS technology) is used, the reduction of CO_2 up to 80 % could be possible.

ULCORED is an advanced DRI shaft furnace using hydrogen and CO from the reaction of shaft furnace off-gas with natural gas as a reducing agent. The aim of the technology is to achieve the highest possible proportion of hydrogen in the reducing gas. The process is also combined with the storage of CO_2 in underground reservoirs. The technology has not yet been operationally tested. It is currently in a pilot technological stage.

The electrolysis is studied on the laboratory scale and in the context of other projects. The electrowinning of metals from an aqueous solution and the molten slag is being tested. Currently, electrolysis of iron ore at $110\,°C$ in an electrolyte – an aqueous solution (ULCOWIN) – produces 3 kg of metallic iron.

There are currently two basic approaches to the reduction of CO_2 emissions. The first approach is to use an alternative reducing agent instead of carbon. One alternative is biomass. The second alternative is hydrogen from transformed coke oven gas (COG). The efforts are focused on further improvement of the efficiency of transformed coke oven gas (COG) using waste heat. The second approach is to reduce the amount of CO_2 escaping into the atmosphere from the top gas by its capture and storage. To increase the effectiveness, the maximum use of waste heat for improved absorption of CO_2 by sorbents is studied.

The most extensive use of biomass in the context of iron production is achieved in Brazil, where the mini blast furnaces use charcoal. The furnaces emit approximately 1790 kg of CO_2 per tonne of pig iron. However, when the charcoal is produced by growing eucalyptus trees on a plantation, in the course of growth, they absorb 3643 t of CO_2 required for the production of the necessary 620–800 kg of charcoal (including the injection) per tonne of iron. A major drawback of operating the mini blast furnace technology using charcoal in Brazil is the considerable devastation of the environment in areas of eucalyptus trees plantations [93].

From an economic point of view and with regard to continuous achievements in research and development, the greatest potential for biomass use is in an integrated process, where it is estimated that CO_2 emissions could be halved. There is also a potential for reducing CO_2 emissions by about 12–15 % in the technology of steel production in electric arc furnaces, where it is feasible to replace the injection of anthracite carbon with carbon from biomass. *Tab. 48* shows the minimum potential for reducing CO_2 emissions using biomass within the technological process of coke, sinter, pig iron and steel production [113].

Tab. 48 Potential for reducing CO_2 emissions using biomass in metallurgy of iron and steel [113]

Process	Substitution by biomass	Decrease of CO_2 emissions	
		t CO_2/t steel	% CO_2
Agglomeration process	50–100 % coke powder substitution (consumption of 45–60 kg of coke powder/t agglomerate)	0.12–0.32	5–15
Coking process	2–10 % coal substitution (consumption of 300–350 kg of coke/t pig iron)	0.02–0.11	1–5
Blast furnace process	100 % injected coal substitution (consumption of 150–200 kg of coal/t pig iron)	0.41–0.55	19–25
Blast furnace process	50–100 % coke nut substitution (consumption of 45 kg of coke nut/t pig iron)	0.08–0.16	3–7
Steelmaking process (BOF)	100 % anthracite substitution (carburiser) (consumption of 0.25 kg of anthracite/t steel)	0.001	0.04
Steelmaking process (EAF)	50–100 % coke substitution (carburiser) (consumption of 12 kg of coke/t steel)	0.019–0.037	3.8–7.5
Steelmaking process (EAF)	50–100 % coke substitution (frother) (consumption of 5 kg of coke/t steel)	0.008–0.016	1.6–3.1

There are two new carbon-free steelmaking processes that have been explored for a long time, Molten Oxide Electrolysis (MOE – electrolysis of Fe oxides) and Hydrogen Flash Smelting (HFS – hydrogen smelting). It is expected that these technologies will contribute to the reduction of CO_2 emissions by 30 % in the near future, and 70 % or more in the long run.

16. Alternative fuels of the future

What will be the fuels of the future? And which of them will be capable of being used in metallurgy? Will they be based on carbon or hydrogen? Will they be more ecological than today? These are the questions being asked by the authors of this publication, and the experience gained in solving the project on the use of biomass in the agglomeration process leads them to believe that at least in the near future, there will be a higher share of renewable sources used. The experts most frequently consider hydrogen. Hydrogen as a fuel of the future has a chance to increase the competition among fuel suppliers and reduce the energy dependence of developed countries on oil imports. It can be produced from renewable sources – biomass, hydro, wind and solar energy – and also from nuclear sources.

Among future technologies focused on fuel, we can include:

- hydrogen and bio-hydrogen production,

- production of bio-methane,

- catalytic liquefaction of biomass,

- converting carbon and hydrocarbon-based waste using plasma processes,

- electro-chemical conversion of biomass,

- production of synthesis gas (conversion of carbon dioxide and water into hydrogen and carbon monoxide) using sunlight,

- development of new types of nuclear reactors based on uranium,

- development of thorium reactors,

- mining of methane hydrates.

Let us now try to predict some potential new alternative fuels of the future. Given that fossil fuels are being gradually depleted, while the human population is growing, it is possible to assume that energy consumption will continue to rise. Therefore, considerable effort is currently dedicated to finding new alternative fuels. These alternative fuels should be an adequate substitute for fossil fuels in terms of energy content, with minimum negative impact on the environment, while their reserves should not be restricted. A very good solution is to store the energy generated from renewable sources in the form of hydrogen.

Hydrogen is the third most abundant element on the Earth and exists primarily as water, and also as a part of organic compounds. In terms of energy use, hydrogen can be classified as a promising alternative fuel, which currently attracts enormous interest both for the reasons of environmental protection and due to the increase in energy consumption and energy prices.

Nowadays, hydrogen is used especially in the chemical industry for the production of ammonia, in the petrochemical industry for crude oil refining, and in the metallurgical industry for production and refining of metals. A small portion of hydrogen is used for the production of methanol as a fuel. There are two ways of using hydrogen as fuel. One is the combustion of hydrogen, and the other is its utilisation in fuel cells. In both cases, the product is ecologically safe water or water vapour. Elemental hydrogen is found naturally only in small quantities and is therefore not a source of energy. For energy purposes, it must be produced by converting the appropriate raw materials using energy. At present, 96 % of hydrogen is produced thermochemically from fossil fuels, especially natural gas. For the future, coal gasification is very promising in terms of obtaining energy gas as well as hydrogen. Among the thermochemical processes for the production of hydrogen, reforming is used the most often used, which is based on fission (reforming) of hydrocarbons from fossil fuels by steam. It is possible to obtain hydrogen with high purity and low emissions by gasification if the separation of CO_2 is ensured. There have been various studies of carbon dioxide separation from the mixture of gases (physical dissolution, chemical absorption, membrane separation or cryogenic separation) so that the resulting product was hydrogen. To produce hydrogen by electrolysis of water, it is necessary to provide a vast amount of cheap electricity. Hence, only a very small proportion of hydrogen (about 5–7 %) is presently produced by electrolysis of water. However, due to the need to save primary energy sources and environmental issues related mainly to the production of CO_2, it can be assumed that the production of hydrogen by electrolysis of water will increasingly expand. Splitting of water molecules by electrolysis can be enhanced thermally (need for extreme temperatures) and chemically (need for expensive catalysts – e.g. palladium). Many substances are capable of splitting water, yet different forms of carbon are economically most viable for large, commercial applications. Various types of coal are the cheapest and the most readily available. Therefore, during the transition towards a hydrogen economy in particular, it is expected that the primary source for hydrogen production will be fossil fuels, mainly coal. In research projects, solutions have been proposed, which can also obtain hydrogen from coal. For instance, dehydrogenation of coal is based on the use of sulfur, which is relatively cheap (it can be recycled during the process) and is a reliable agent for dehydrogenation of organic compounds forming H_2S. The reactions take place at relatively low temperatures in the range of 230–650 °C. The high carbon solid mixture of dehydrogenation by-product can be a raw material for the production of carbon materials. The solid carbonaceous product has properties of high-quality metallurgical coke, although it has been produced at temperatures below the conventional coke ovens. A research on the more efficient use of hydrogen in metallurgical reduction processes is currently being conducted [126].

Among future alternative fuels, a variety of products produced within the technologies of energy recovery from waste can also be included – e.g. sludge and solid municipal waste, synthetic waste, etc. The already known technologies of gasification, pyrolysis or liquefaction can be employed.

Although nuclear energy has been used for decades, newer and safer nuclear reactors are constantly being developed. The latest Generation III reactors have layers of passive protection, which are designed to prevent melting of the overheated core of the nuclear reactor. Such design of the reactor is about to be implemented in the near future in many nuclear power plants.

Some experts from the nuclear energy sector have already announced the development of Generation IV reactors, including thorium reactor. In this case, liquid thorium would replace solid uranium used in existing power plants. Such change would mean that the meltdown would be virtually impossible. These types of reactors have two major safety benefits. Their liquid fuel is under much lower pressure than solid fuel. This hugely reduces the likelihood of an accident, such as a hydrogen explosion. In case of power failure, frozen salt in the reactor melts, and liquid fuel drains into tanks, where it solidifies, and fission reaction stops. Thorium provides, in addition to the safety, also other strategic advantages. The need for huge cooling towers is dramatically reduced – thorium power plant would, therefore, be much smaller in size, as well as production capacity.

At present, there is a research project in the field of thermonuclear energy whereby the energy would be produced from the fuel consisting of a combination of heavy hydrogen (deuterium) and tritium.

Oil shale and gas obtained from shale appears to be a certain alternative to the present fossil fuels. Deposits of shale are found in Colorado, Utah and Wyoming. Although shale constitutes a certain alternative, it is not quite a perfect energy source. The process of obtaining oil and the gas in this manner is economically and environmentally challenging. Therefore, further research will be necessary in this field.

17. Conclusion

Probably for a long time to come, the world will be divided into groups arguing about the global warming, the greenhouse effect and carbon dioxide, as well as about the contribution of the humankind to its concentration in the atmosphere. We will argue which human activity has the most negative impact on the environment. One thing, however, is indisputable. There are more people on Earth every year, which is very closely related to the systematic removal of vegetation from the Earth's surface and rising energy consumption. Large quantities of substances more dangerous than carbon dioxide are emitted into the atmosphere. These substances are able to remain in the atmosphere not only for months but for years. They are carried over vast distances by airflow and cause environmental problems far away from the sources which emitted them into the atmosphere. The question of emissions is already not only about local sources of pollution but is now a global issue. It is important all the countries address the environmental issues equally responsibly. From the global perspective, the reduction of emissions in the EU countries is almost negligible against the continuous increase of emissions in countries such as China, India, Brazil, etc. Nevertheless, this effort in the EU has its significance, and it has managed to locally reduce emissions in the region for the last ten years. However, we must seek global solutions for emission reduction in every industry, metallurgy being no exception.

Also, this publication has the ambition to contribute to the expanding knowledge on the use of traditional and alternative fuels in metallurgy. Although it is possible to produce a part of the energy from renewable biomass sources in the metallurgy industry as well, it is essential to responsibly and comprehensively assess the main advantages and drawbacks of this alternative source of energy. Replacement of fossil fuels with renewable energy sources cannot rely only on increasing the production of biofuels. It appears that such production and subsequent processing of biomass products has a substantial devastating effect on the environment in some countries (e.g. in Brazil, Central and South Africa). Nonetheless, it is crucial to search and develop new directions in the use of alternative fuels for the metallurgical sector. The use of biomass within the high-temperature metallurgical processes contributes to the overall lower production of emissions compared to fossil fuels. It is the most notable in emissions of sulfur and nitrogen. In the case of carbon compounds (mainly CO_2, CO and total organic carbon), there are too many influencing parameters in the high-temperature processes. Yet, there is still a room for optimisation of metallurgical technologies using biomass. If Slovakia wants to respond to the changing global economic and environmental conditions, it should invest in the research and development of biomass in the metallurgical technologies as well. Only the biomass, whose use does not compromise the long-term regenerative potential and ecological stability of the site of its origin, should have the status of the sustainably renewable source. The use of waste biomass seems to be the best in this respect. The research of such biomass within the metallurgical technologies in Slovakia is only minimal in comparison with the developed countries of the world.

Of course, it is impossible to argue that we will manage without fossil fuels in the future. Historically, fossil fuels always had, have and will have their application. Also, the production of biofuels is often associated with the high energy demand, which is largely satisfied by the traditional fossil fuels. Therefore, the responsible approach to energy saving and economical use of technologies not only on the basis of renewable energy sources is necessary. Not only the conscious behaviour of producers and users of energy, but also new technologies capable of utilising renewable energy sources can contribute to the energy savings.

The research of biomass in the conditions of the Slovak Republic has hitherto been carried out mainly in the energy industry. This publication entitled "Biomass and Carbon Fuels in Metallurgy" demonstrates that the energy potential of biomass can also be successfully used in the pyrometallurgical processes. The near future of the metallurgical industry in our country and in the world will show whether or not it will happen.

In the introduction to this publication, a sentence was written, which is its synonym. As the authors are writing these last sentences, they do not need to change anything in its content, for it reflects their relationship to the use of alternative fuels – mainly the biomass: "By effective use of even a small amount of wood or plant matter from secondary materials in metallurgy, we gain not only the benefit of saving a certain amount of conventional fossil fuels, but we mainly achieve the potential and experience in operating the metallurgical production technologies using alternative fuels".

List of figures

List of tables

References

[1] Jandačka J., Malcho M.: Biomasa ako zdroj energie. GEORG. Žilina. máj 2007. ISBN 978-80-969161-4-6

[2] Hartmann H., Thuneke K., Holdrich A., Rozmann P.: Handbuch bioenergie - kleinanlagen. FNR Mit Förderung des Bundesministeriums fur Verbraucherschutz. Ernährung und Landwirtschaft. 2003. ISBN 3-00-011041-0

[3] Jandačka J., Malcho M., Mikulík M.: Biomasa ako zdroj energie. Potenciál. druhy. bilancia a vlastnosti palív. GEORG. Žilina. január 2007. ISBN 978-80-969161-3-9

[4] Fosilní paliva. Dostupné z internetu: http://www.fospaliva.wz.cz/index.htm

[5] Lintnerová O.: Geológia kaustobiolitov. Uhlie a uhľovodíky. Univerzita Komenského. Bratilava. 2009. ISBN 978-80-223-2623-0

[6] http://www.worldcoal.org/coal/

[7] Findorák R., Mašlejová A., Legemza J., Kromková Z.: Petrografia pre hutnícku prax. Skriptá. TU v Košiciach. 2014

[8] Classification of coals by rank. Standard D-388.: American society for testing materials. Section 5. 1983

[9] http://en.wikipedia.org/wiki/Coal

[10] Crelling J., Russel D.: Principles and applications of coal petrology. Indiana University at Bloomington. United States. 1980

[11] Van Krevelen., D. Coal. Elsevier scientific publishing. New York. 1961

[12] Suarez-Ruiz I., Crelling. J.: Applied coal petrology. the role of petrology in coal utilization. Oxford. 2007

[13] Fallot A., Saint-André L., Laclau J.P., Nouvellon Y., Marsden C., Le Maire G., Silva T., Piketty M.G., Hamel O., Bouillet J.P.: Biomass sustainability. availability and productivity. Proceedings of the 4th Ulcos seminar. 1.-2. October 2008

[14] http://www.biomassenergycentre.org.uk

[15] Ogura S. et al.: Environmental conservation and energy saving activities in JFE Steel. JFE Technical Report. 2014

[16] Roubíček V., Buchtele J.: Uhlí. zdroje. procesy. užití. Montanex. 2002. ISBN 80 7225-063-9

[17] Legemza J.: Habilitačná práca. TU v Košiciach. 2009

[18] International Iron and Steel Institute. http://www.worldsteel.org/

[19] Integrovaná prevence a omezování znečištění (IPPC): Referenční dokument k aplikování nejlepších dostupných technik (BAT). Dostupné z internetu http://www.ippc.cz

[20] ULCOS. the European initeiative for CO_2-lean steelmaking: Mefos. 2006

[21] Worell E. et al.: Energy efficiency and carbon dioxide emisions reduction opportunities in the U.S. iron and steel sector. 1999

[22] Coal information: International Energy Agency. 2011

[23] World coal resources. PANORAMA. 2010

[24] Fukuda S.: Biomass characterization. KMUTT. 2012

[25] Ochodek T., Koloničný J., Janásek P.: Potenciál biomasy. druhy. bilance a vlastnosti palív z biomasy. VŠB – TU Ostrava. 2006

[26] Frőhlichová M., Legemza J., Findorák R., Mašlejová A.: Biomass as a source of energy in iron ore agglomerate production process. 2014. In: Archives of Metallurgy and Materials. Vol. 59. no. 2 (2014). p. 815-820. ISSN 2300-1909

[27] Kucková J.: Hutníctvo železa – výroba cokeu. Skriptá. TU Košice. HF. KMŽaZ. 2003

[28] Taylor G.H., Teichmüller M., Davis A., Diessel C. F. K., Littke R., Robert P.: Organic petrology. Berlin. Stuttgart. Borntraeger. 1998. ISBN 3-443-01036-9

[29] The new vitrinite classification (ICCP System 1994): Fuel . Vol. 77. No 5. p. 349 – 358. 1998

[30] Košina M.: Mikropetrografické metody v hodnocení černých uhlí. Ústav pro výzkum a využití paliv. Praha. 1979

[31] Myrvågnes V.: Analyses and characterization of fossil carbonaceous materials for silicon production. Thesis for the degree philosophiae doctor. Norwegian University of Science and Technology. Trondheim. January 2008

[32] The new inertinite classification (ICCP System 1994); Fuel. Vol.80. p. 459 – 471. 2001

[33] Gan M., Fan X., Chen X., Ji Z., Lv W., Wang Y., Yu Z., Jiang T.: Reduction of pollutant emission in iron ore sintering process by applying biomass fuels. ISIJ International. Vol. 52. No. 9. p. 1574 – 1578. 2012

[34] Majerčák Š.: Hutníctvo surového železa. II. diel - Agglomeration. VŠT Košice. 1979

[35] Majerčák Š. Majerčáková. A.: Vysokopecná vsádzka. Praha. SNTL. 1986

[36] Legemza J.: Doktorandská dizertačná práca. TU v Košiciach. 1998

[37] Findorák R.: Doktorandská dizertačná práca. TU v Košiciach. 2010

[38] Brož L.:Teoretické základy výroby železa. Praha. SNTL/Alfa. 1975

[39] Usamentiaga R. et al.: Monitoring sintering burn-through point using infrared thermography. Sensors. Vol. 13. 2013. p. 10288-10305. DOI: 10.3390/s130810287

[40] Machida S., Sato H., Takeda K.: Development of the process for producing pre-reduced agglomerates. JFE Technical Report. Vol. 1. 2009. No. 13. p. 7-13

[41] Heinänen K: Mineralogy and metallurgical properties of iron ore sinter based on magnetite fines. Thesis for the Degree of Doctor of Philosophy. University of Helsinki. Finland. 1993

[42] Jonsson C.: Deposit formation in the Grate-Kiln process. Department of Engineering Sciences and Mathematics Luleå University of Technology. Luleå. Sweden. 2013

[43] T. van den Berg: An assessment of the production of fine material in iron ore sinter. Dissertation. Department of Materials Science and Metallurgical Engineering. Pretoria. 2008

[44] Avila C.: Predicting self-oxidation of coals and coal/biomass blends using thermal and optical methods. Thesis submitted to The University of Nottingham for the degree of Doctor of Philosophy. May 2012

[45] Mohamed F. M., Hussiny N. A. El. Shalabi M. E. H.: Granulation of coke breeze fine for using in the sintering process. Science of Sintering. 42 (2010) 193-202. DOI: 10.2298/SOS1002193M

[46] Qu Y.: Experimental study of the melting and reduction behaviour of ore used in the HIsarna process. PhD thesis. Department of Materials Science and Engineering of Delft University of Technology in the Netherlands. 2013

[47] Skotland Ch. H.: Measurement of temperature conditions in grate zone of a 1 MW wood-pellets boiler fired with high ash content wood-pellets. Norwegian University of Science and Technology. June 2009

[48] Majerčák. Š., Karwan T.:Theory of sintering fine materials. Košice. 1998

[49] Paananen T.: The effect of minor oxide components on reduction of iron ore agglomerates. Doctoral Thesis University of Oulu Department of Process and Environmental Engineering Laboratory of Process Metallurgy. 2013

[50] Legemza J., Frőhlichová M., Findorák R., Bakaj F.: Emissions CO and CO_2 in the sintering process. In: SGEM 2010 : 10th international multidisciplinary scientific geoconference : conference proceedings : Volume 2 : 20-26. June. 2010. Bulgaria. Sofia. SGEM. p. 567-572. ISBN 978-954-91818-1-4

[51] Legemza J., Frőhlichová M., Findorák R., Bakaj F.: The process of simulating the agglomerate laboratory production under laboratory conditions. In: Acta Metallurgica Slovaca: Conference : Iron and Steelmaking. 20. - 22.10.2010. Tatranská Lomnica. TU. 2010. Roč. 1. č. 4 (2010). s. 70-75. ISSN 1338-1660

[52] Mutombo N.: Study of sinter reactions when fine iron ore is replaced with coarse ore. using an infrared furnace and sinter pot tests. Dissertation. Department of Materials Science and Metallurgical Engineering. Pretoria. 2011

[53] Legemza J., Findorák R., Baricová D., Bakaj F.: Optimalizácia druhov mangánových rúd pre výrobu mangánových ferozliatin v OFZ. a. s. bez použitia aglomerátu. TU v Košiciach. 2012

[54] Legemza J., Frőhlichová M., Findorák R., Bakaj F.: Laboratórne overenie výroby Mn of agglomerate – ako vsádzky pre výrobu FeSiMn v OFZ. a.s. TU v Košiciach. 2014

[55] Semanová Z., Legemza J.: Analysis and use of Mn ore fines. 2014. In: Acta Metallurgica Slovaca. Roč. 20. č. 4 (2014). s. 410-417. ISSN 1335-1532

[56] Semanová Z., Legemza J.: Study of properties of Mn ore fines and possibilities their utilization in the production of FeSiMn. In: Acta Metallurgica Slovaca. Roč. 21. č. 1 (2015). s. 68-77. ISSN 1335-1532

[57] Legemza J., Findorák R., Sajková M.: The thermodynamic study of dust and sludge from iron and steel industry. In: Prace instytutu metalurgii zelaza. Tom 58. no. 4 (2006). p. 125-127. ISSN 0137-9941

[58] Legemza J., Majerčák Š.: Štúdium medzifázového rozhrania tavenina - oxid v metalizovaných aglomerátoch. In: Acta Metallurgica Slovaca. Roč. 3. special issue 1/2 (1997). s. 306-312. ISSN 1335-1532

[59] Legemza J.: The possibilities of utilitizing of dust and sludge from steel industry. In: Acta Metallurgica Slovaca. Roč. 10. č. 2 (2004). s. 80-87. ISSN 1335-1535

[60] Legemza J.: Termodynamické štúdium oceliarenského prachu a kalu. In: Acta Metallurgica Slovaca. Roč. 10. č. 3 mimoriadne (2004). s. 573-576. ISSN 1335-1532

[61] Bennett P., Fukushima T.: Impact of PCI coal quality on blast furnace operations. 12th ICCS - November 2003

[62] Kret J.: Požiadavky na akosť vysokopecného cokeu pre moderné technológie výroby železa. Acta Metalurgica Slovaca. No 4 (1997). p. 325-331

[63] Olejár M., Mašlejová A., Varga R., Nemčovský P.: PCI technology evaluation. Hutnicke listy. Vol.XLIX. No.6. 1995. p.6-12. ISSN 0018-8069

[64] Kret J.: Vliv alkálií na výrobu surového železa ve vysoké peci. Hutnické listy. 2000. č.4. s.10-14. ISSN 0018-8069

[65] Lundgren M.: Blast furnace coke properties and the influence on off-gas dust. Licentiate Thesis. Luleå University of Technology. Luleå. Sweden 2010

[66] http://www.jap.cz/grafit/nauhlicovadla/standardni-nauhlicovadla/

[67] http://www.carbon-vum.sk/

[68] Maki A., Ariyama T.: Ironmaking technologies contributing to the steel industry in the 21st century. In: NKK Technical Review. 2003. No.88. p. 10-17.

[69] Mikulík M., Jandačka J.: Postupy správneho vykurovania. ERDF. 2013

[70] Kasai E. et al: Macroscopic behaviours of dioxins in the iron ore sintering plants. In: ISIJ international. Vol.41. 2001. No. 1. p. 86-92.

[71] Fisher R., Fray T.: Investigation of the Formation of Dioxins in the Sintering Process. In: 2nd International Congress on the Science and Technology of Ironmaking Conjunction with the 57th Ironmaking Conference of Iron and Steel Society. C-Toronto. 1998

[72] Kim J.R., Lee K.J., Hur N.S.: Improvement of sinter plant stack emissions at Kwangyang Works. Posco. In: 2nd International Congress on the Science and Technology of Ironmaking Conjunction with the 57th Ironmaking Conference of Iron and Steel Society. C-Toronto. 1998

[73] www.siemens-vai.com: Maximized emission reduction of sintering – Simetal [CIS] Meros plant. Linz: 2008. Dostupné na internete: <https://www.industry.siemens.com/datapool/industry/industrysolutions/metals/simetal/en/SIMETAL-MEROS-plant-en.pdf>.

[74] Tan P., Neuschütz D.: Study on polychlorinated dibenzo-p-dioxin/furan formation in iron ore sintering process. In: Metallurgical and Materials Transactions. Vol.35. 2004. No.5. p. 983-991.

[75] Aries E., et al: PCDD/F and "dioxin-like" PCB emissions from iron ore sintering plants in the UK. In: Chemosphere. Vo.65. 2006. No.9. p.1470-1480

[76] Wang L. Ch., et al: Emissions of polychlorinated dibenzo-p-dioxins and dibenzofurans from stack flue gases of sinter plants. In: Chemosphere. Vol. 50. 2003. No.9. p. 1123-1129.

[77] Malaťák J., Vaculík P.: Biomasa pro výrobu energie. Praha. Česká zemědělská univerzita v Praze. 2008

[78] Kára J., a kol.: Využití biomasy pro energetické účely. Praha. CEA. 1997

[79] Gan M., et al.: Reduction of pollutant emission in iron ore sintering process by applying biomass fuels. In: ISIJ International. Vol. 52. 2012. No. 9. p. 1574 – 1578

[80] Llorente M.J. F., García J.E C.: Comparing methods for predicting the sintering of biomass ash in combustion. In: Fuel. Vol. 84. 2005. No.14. p. 1893 – 1900

[81] Chen X.L., Wang Y.Z., Liu X., Liang Q.F.: Effect of biomass addition on ash fusion characteristic of coal with high ash melting point. 5th Internacional Freiberg Conference on IGCC. Germany. 2012

[82] Mabizela P. S., Meyer E.L., Mamphweli N.S.: Thermal characterization of various biomass materials for co-gasification with coal. Institute of Technology University of Fort Hare. Private Bag 1314. Alice. 5700. South Africa

[83] Pustějovská P., Jursová S.: Process engineering in iron production. Chemical and process Engineering. Vol.34. 2013. No. 1. p. 63-76. DOI: 10.2478/cpe-2013-0006

[84] Lovel R., Vining K., Dell'Amico M.: Iron ore sintering with charcoal. In: Mineral Processing and Extractive Metallurgy. Vol. 116. 2007. No. 2. p. 85 – 92

[85] Kawaguchi T., Hara M.: Utilization of biomass for iron ore sintering. In: ISIJ International. Vol.53. 2013. No.9. p.1599-1606

[86] Arromdee P., Kuprianov V. I.: A comparative study on combustion of sunflower shells in bubbling and swirling fluidized-bed combustors with a cone-shaped bed. In: Chemical Engineering and Processing: Process Intensification.Vol. 62. 2012. p. 26 – 38

[87] Biagini E., Cioni M., Tognotti L.: Development and characterization of a lab-scale entrained flow reactor for testing biomass fuels. In: Fuel. Vol. 84. 2005. No. 12 – 13. p. 1524 – 1534

[88] Doulati A.F. et al.: Adsorption of direct Red 80 dye from aqueous solution onto almond shells: Effect of pH. initeial concentration and shell type. In: Journal of Hazardous Materials. Vol. 15. 2008. No. 2 – 3. p. 730 – 737

[89] Murakami K., Sugawara K., Kawaguchi T.: Analysis of combustion rate of various carbon materials for iron ore sintering process. In: ISIJ International. Vol. 53. 2013. No.9. p. 1580–1587

[90] Norgate T., Langberg D.: Environmental and economic aspects of charcoal use in steelmaking. ISIJ International. 2009. No. 49. p. 587-595

[91] Zandi M., et al: Biomass for iron ore sintering. In: Minerals Engineering. Vol. 22. 2010. No.14. p. 1139 – 1145

[92] Thomas S., et al: Laboratory evaluation of biomass usage for coke and sinter production. In: METEC. InSteel. 2011

[93] Ooi T. Ch., et al: The effect of charcoal combustion on iron-ore sintering performance and emission of persistent organic pollutants. In: Combustion and Flame. Vol.158. 2011. No.5. p. 979-987

[94] Ooi T. C., et al.: The study of sunflower seed husks as a fuel in the iron ore sintering process. In: Minerals Engineering. Vol. 21. 2008. p. 167 – 177

[95] Lovel R., Vining K., Dell'Amico M.: The influence of fuel reactivity on iron ore sintering. In: ISIJ International. Vol. 49. 2009. No. 2. p. 195 – 202

[96] Fan X., Ji Z., Gan M., Chen X., Li W., Yu Z.: Strengthening refractory iron ore sintering with biomass fuel. Florida. 3rd International symposium on high-temperature metallurgical processing – TSM 2012. p. 357 – 364

[97] Mašlejová A.: Evaluation of iron ore sinter structure using a various types of biomass. In: 13th SGEM GeoConference on Science and Technologies In Geology. Albena (BULGARIA). Vol. 2. 2013. p. 581-588. DOI:10.5593/SGEM2013/BA1.V2/S04.007

[98] Mašlejová A.: Utilization of Biomass in Ironmaking. 22th International Conference Metallurgy and Materials – METAL 2013. May 2013 Brno. Czech Republic. p. 38

[99] Thomas S., McKnight S.J., Serrano E., Mašlejová A., Želinský R., Tomáš J., Vlašič P.: Laboratory evaluation of biomass usage for coke and sinter production. Alternative Fuels in Iron- and Steelmaking. Proc. METEC InSteelCon. 2011

[100] Legemza J., Frőhlichová M., Findorák R.:Termodynamické štúdium využitia biomasy v rámci aglomeračného procesu. In: Wybrane zagadnienia energo-fizyczne w produkcji stali : Praca zbiorowa pod redakcja naukowa dra inž. Artura Hutnego. - Czestochowa: Wydawnitwo Wydzialu Inzynierii Procesowej. Materialowej i Fizyki Stosowanej Politechniki Czestochowskiej. 2013 p. 5-11. ISBN 978-83-63989-11-8

[101] Legemza J., Frőhlichová M., Findorák R.: Štúdium použitia biomasy v rámci spekania železonosných materiálov vo svete. In: Wybrane zagadnienia energo-fizyczne w produkcji stali: Praca zbiorowa pod redakcja naukowa dra inž. Artura Hutnego. - Czestochowa : Wydawnicwo Wydzialu Inzynierii Procesowej. Materialowej i Fizyki Stosowanej Politechniki Czestochowskiej . 2013 p. 12-21. ISBN 978-83-63989-11-8

[102] Legemza J., Frőhlichová M., Findorák R.: Thermodynamic study of utilization of charcoal in the iron-ore sintering process. In: SGEM 2013: 13th International Multidisciplinary Scientific Geoconference Science and Technologies in Geology. Exploration and Mining: conference proceedings: Vol. 2. 16-22. June. 2013. Albena. Bulgaria. p. 649-656. ISBN 978-954-91818-8-3

[103] Legemza J., Frőhlichová M., Findorák R.: Thermodynamic study of utilization of sawdust from pine-wood in the iron-ore sintering process. In: SGEM 2013: 13th International Multidisciplinary Scientific Geoconference Science and Technologies in Geology. Exploration and Mining: conference proceedings. Vol. 2. 16-22. June. 2013. Albena. Bulgaria. p. 657-663. ISBN 978-954-91818-8-3

[104] Frőhlichová M., Findorák R., Legemza J.: The possibility of using of biomass in the Agglomeration process. In: SGEM 2014 : 14th International Multidisciplinary Scientific Geoconference: Science and Technologies in Geology. Exploration and Mining: conference proceedings. Vol. 3. 17-26 June. 2014. Albena. Bulgaria. p. 981-986. ISBN 978-619-7105-09-4

[105] Findorák R., Frőhlichová M., Legemza J.: The study of saw-dust addition on iron-ore sintering performance. In: Acta Metallurgica Slovaca - Conference: Iron and Steelmaking : 23. international conference. Vysoké Tatry. Štrbské pleso. TU. 2014 s. 23-30. ISSN 1338-1660

[106] Frőhlichová M., Findorák R., Legemza J.: Structural Analysis of Sinter with Titanium Addition. In: Archives of Metallurgy and Materials. Vol. 58. No. 1. 2013. p. 179-185. ISSN 1733-3490

[107] Frőhlichová M., Legemza J., Findorák R., Bakaj F.: The process of simulating the iron ores agglomerate production under laboratory conditions . In: Hutník. Vol. 77. No. 9. 2010. p. 450-454. ISSN 1230-3534

[108] Frőhlichová M., Legemza J., Sajková M., Findorák R.: Decrease of energy load and emissions of CO and CO_2 at the production of pig iron. In: Research and Development Projects. Košice. HF TU. 2007. p. 15-16. ISBN 9788080738303

[109] Findorák R., Frőhlichová M., Legemza J.: The effect of charcoal addition on iron-ore sintering emision. In: SGEM 2013: 13th International Multidisciplinary Scientific Geoconference Science and Technologies in Geology. Exploration and Mining: conference proceedings. Vol. 2. 16-22 June. 2013. Albena. Bulgaria. p. 629-635. ISBN 978-954-91818-8-3

[110] Findorák R., Frőhlichová M., Legemza J.: The effect of charcoal addition on iron-ore sintering performance. In: SGEM 2013: 13th International Multidisciplinary Scientific Geoconference Science and Technologies in Geology. Exploration and Mining: conference proceedings. Vol. 2. 16-22 June. 2013. Albena. Bulgaria. p. 637-642. ISBN 978-954-91818-8-3

[111] Findorák R., Frőhlichová M., Legemza J.: The effect of saw-dust addition from pine and oak wood on iron-ore sintering performance. In: SGEM 2014: 14th International Multidisciplinary Scientific Geoconference: Science and Technologies in Geology. Exploration and Mining: conference proceedings. Vol. 3. 17-26. June. 2014. Albena. Bulgaria. p. 973-980. ISBN 978-619-7105-09-4

[112] Legemza J., Frőhlichová M., Findorák R.: The thermovision measurement of temperature in the iron-ore sintering process with the biomass. In: Acta Metallurgica Slovaca - Conference: Iron and Steelmaking: 23. international conference: Vysoké Tatry. Štrbské pleso. TU. Vol. 4. 2014. p. 56-65. ISSN 1338-1660

[113] Mathieson J., Rogers H., Somerville M. A., Jahanshahi S., Ridgeway P.: Potential for the use of biomass in the iron and steel industry. Port Kembla. Newcastle

[114] Mežibrický R., Frőhlichová M., Mašlejová A.: Fázové zloženie aglomerátov vyrobených s náhradou prachového paliva. In: Iron and steelmaking: 24. international scinetific conference. conference proceedings: Horní Bečva. Beskydy. Česká republika. VŠB-TU. 2014. s. 17-20. ISBN 978-80-248-3627-0

[115] Suopajärvi H.: Bioreducer – possibilities of bio-based materials in reduction applications. EU material

[116] New blast furnace process. rtd-steel-coal@ec.europa.eu

[117] Zhang W.: Automotive fuels from biomass via gasification. Fuel Proc Technol 91(8). 2010. p. 866–876

[118] Costa R., Wagner D., Patisson F: Modelling a new. low CO_2 emissions. hydrogen steelmaking process. Journal of CleanerProduction. 2013. No. 46. p. 27–35 http://dx.doi.org/10.1016/j.jclepro.2012.07.045

[119] Minoru A., Masahiko K., Minoru K., Yasuhiro F., Kaneo T.: Establishment of advanced recycling technology for waste plastics in blast furnace. JFE Technical report. No. 13. 2009

[120] https://asbury.com

[121] Jandačka J., Malcho M., Mikulík M.: Ekologické aspekty zámeny fosílnych palív za biomasu. Žilina. 2008. 228 s., ISBN 978-80-969595-5-6

[122] Ochodek T., Koloničný J., Branc M.: Ekologické aspekty záměny fosilních paliv za biomasu. VŠB TU Ostrava. 2007. ISBN 978-80-248-1595-4

[123] Pustějovská. P., Jursová. S., Brožová. S.: Determination of kinetic constants from tests of reducibility and their application for modelling in metallurgy. Journal of the Chemical Society of Pakistan. Volume 35. Issue 3. June 2013. p. 565-569. ISSN 0253-5106

[124] Pustějovská P., Brožová S., Jursová S.: Environmental Benefits of Coke Consumption Decrease. In 19th International Conference on Metallurgy and Materials: In Metal 2010. Rožnov pod Radhoštěm. 18. – 20. 5. 2010. Tanger. spol. s. r. o., Ostrava. s. 79-83. ISBN 978-80-87294-17-8

[125] Brožová S., Kret J.: Using of plasma technologies for processing of various wastes. Hutnik. Wiadomości hutnicze. 2006. Vol. LXXIII. No. 7. p. 343-247. ISSN 1230-3534

[126] Bilik J., Pustějovská P., Brožová S., Jursová S.: Efficiency of hydrogen utilization in reduction processes in ferrous metallurgy. Scientia Iranica. 20 (2013). pp. 337-342 DOI information: 10.1016/j.scient. 2012.12.028

[127] Leško J., Hudák J., Semanová Z.: Impact of biofuel in agglomeration process on production of pollutants, 2017. In: Science of Sintering. Vol. 49, no. 2 (2017), p. 159-166. - ISSN 0350-820X

[128] Hudák J., Leško J., Semanová Z., Legemza J., Frőhlichová M., Findorák R.: Evaluation of granulability and absorptivity of biomass in agglomeration mixture, 2017. In: Journal of Central South University. Vol. 24, no. 10 (2017), p. 2260-2265. - ISSN 2095-2899

[129] Leško J., Legemza J., Hudák J., Frőhlichová M., Findorák R.: Influence of pine and oak wood sawdust addition in the production of iron ore sinter on the environment / Jaroslav Leško ... [et al.] - 2015. In: SGEM 2015. - Sofia : STEF92 Technology Ltd., 2015 P. 629-636. - ISBN 978-619-7105-31-5

[130] Gaurav Jha, S. Soren: Study on applicability of biomass in iron ore sintering process, In: Renewable and Sustainable Energy Reviews, Vol. 80, 2017, p. 339–407

[131] Ramchandra Pode. Potential applications of rice husk ash waste from rice husk biomass power plant. Renew Sustain Energy Rev 2016;53:1468–85

[132] Chen WH, Lin BJ, Huang MY, Chang JS. Thermochemical conversion of microalgal biomass into biofuels: a review. Bioresour Technol 2015;184:314–27

[133] Vassilev SV, Vassileva CG, Vassilev VS. Advantages and disadvantages of composition and properties of biomass in comparison with coal: an overview. Fuel 2015;158:330–50.

[134] Zhao JP, Loo CE, Dukino RD. Modelling fuel combustion in iron ore sintering. Combust Flame 2015;162:1019–34

[135] Abreu GC, de Carvalho JA, Jr, da Silva BEC, Pedrini RH. Operational and environmental assessment on the use of charcoal in iron ore sinter production. J Clean Prod 2015;101:387–94

[136] Min G, Xiao-hui F, Tao J, Xu-ling C, Zhi-yuan Y, Zhi-yun J. Fundamental study on iron ore sintering new process of flue gas recirculation together with using bio-char as fuel. J Cent South Univ 2014;21:4109–14

[137] Lu L, Adam M, Kilburn M, Hapugoda S, Somerville M, Jahanshahi S, Mathieson J Gordon. Substitution of charcoal for coke breeze in iron ore sintering. ISIJ Int 2013;53(9):1607–16